The l
in Hollywood

The Limits of #MeToo in Hollywood

Gender and Power in the Entertainment Industry

MARGARET TALLY

McFarland & Company, Inc., Publishers
Jefferson, North Carolina

Library of Congress Cataloguing-in-Publication Data

Names: Tally, Margaret, 1960– author.
Title: The limits of #Metoo in Hollywood : gender and power in
the entertainment industry / Margaret Tally.
Description: Jefferson : McFarland & Company, Inc.,
Publishers, 2021. | Includes bibliographical references and index.
Identifiers: LCCN 2021021162 |
ISBN 9781476684956 (paperback : acid free paper) ∞
ISBN 9781476642925 (ebook)
Subjects: LCSH: Women in the motion picture industry—United
States. | Women in motion pictures. | Women on television. |
Sex discrimination in employment—United States. | MeToo
movement. | BISAC: PERFORMING ARTS / Film / General |
SOCIAL SCIENCE / Women's Studies
Classification: LCC PN1995.9.W6 T35 2021 | DDC 384/.80810973—dc23
LC record available at https://lccn.loc.gov/2021021162

British Library cataloguing data are available

ISBN (print) 978-1-4766-8495-6
ISBN (ebook) 978-1-4766-4292-5

Front cover image © 2021 Shutterstock

Printed in the United States of America

McFarland & Company, Inc., Publishers
Box 611, Jefferson, North Carolina 28640
www.mcfarlandpub.com

Table of Contents

Acknowledgments

Many people were instrumental in helping me write this book. My long-time collaborator and role model, Betty Kaklamandiou, unfailingly read portions of this manuscript and made the English far smoother than I ever could. My husband, Bill, not only helped with making editorial insights about the manuscript, but more importantly, was a true partner and immeasurable source of support in helping me recover from two broken ankles (!) which laid me up and indirectly helped me to focus on this project. I also want to thank my dean, Nathan Gonyea, and my provost, Meg Benke, at SUNY Empire State College, for being so supportive of my scholarly research, as well as our president, Jim Malatras, who supported me for SUNY Distinguished Professor in the middle of writing this book, which also served to encourage me to continue this project. As the COVID-19 pandemic engulfed us all, my daughters Serena and Lila, in their mid-twenties, came home and also served as supports and constant critics of contemporary media representations of women, and helped me see how important this kind of critical media scholarship can be especially for the younger generations.

Finally, I would like to acknowledge and dedicate this book to my mom, Marjorie Heide, who passed away last year but who was one of the first people to show me how important it was to connect one's life works to helping others and empowering women in particular to find their voices and speak their truth. This book is an effort to do just that, for the millions of women who have experienced sexual harassment and assault and the desire to ensure that their voices will be heard in all walks of life, including our media representations.

Preface

The #MeToo movement gained widespread attention after Hollywood producer Harvey Weinstein was accused of sexual assault and harassment. While it is possible to date its origins to 2006, when an activist named Tatiana Burke heard of a 13-year-old girl who had been assaulted and tried to bring media attention to it, the movement came to national attention on October 15, 2017, when actress Alyssa Milano told her followers on Twitter that "[i]f you've been sexually harassed or assaulted write 'me too' as a reply to this tweet," at which point she received over 66,000 replies. What followed was a huge wave of similar stories that flooded social media, accompanied by the "Me too" hashtag. Soon after, high-profile journalists Ronan Farrow and Jodi Kantor published investigative pieces about sexual misconduct in *The New Yorker* and the *New York Times,* respectively, encouraging even more victims to speak up.

A central theme of the #MeToo movement is the importance of survival and healing in the aftermath of sexual harassment and assault. Another concerns the importance of giving survivors a voice to speak out about their experiences, and in doing so, help many others understand that they are not alone. It is not surprising that, as the #MeToo movement has encouraged women to speak out and at the same time shone a critical light on the power of male film and TV executives, both female and male media producers would seek to tell stories that reflect a new awareness of gender and power relations. And indeed, the past three years have seen a huge influx of stories and characters in film and television that refer to themes raised by the #MeToo movement. This has coincided with a broader critique of structural gender inequalities in the entertainment industry that allowed few women into executive suites and in creative roles. In response, while the absolute numbers are still low, more women have been given creative license to tell stories in film and television—including stories about the realities of sex, gender and power in the workplace. So far however, few works have sought to map the #MeToo movement's impact on the changing representations of women and power in entertainment media; this work aims to begin this task.

This book analyzes how women are portrayed with a particular focus on streaming, cable and net television series as well as a comparative perspective from outside the U.S. In addition to the storylines and characters themselves, this manuscript also explores the interrelated social, political and economic forces that have come together and have coincided with the #MeToo movement to produce a high number of quality series and films. While there have been many inroads, then, in terms of storytelling on film and television that reflect #MeToo themes, there has been less work done in terms of changing the structure of Hollywood to become more inclusive of women as creatives, showrunners, writers and executives. I became interested in this subject after seeing many stories on film and television that drew on the #MeToo movement themes, but at the same time, the overall number of women who were directors of films, or even film and television shows that had female characters, had not changed and I wanted to understand why this was the case. I also wanted to understand why the backlash in the media over whether the #MeToo movement had gone too far was happening. As well, why were many of the men who had been accused of misconduct being brought back into the entertainment industry after a relatively short amount of time? While I had begun to read some really interesting research on issues such as the rhetoric of the #MeToo movement or looking at how it has changed gender dynamics in certain industries, I was curious what the impact had been and ultimately will be for Hollywood, which is arguably where it first began to be voiced in a public way that rose to the level of attention it received in the media. To this end, I studied not only the film and television shows that were coming out post–2017 and that had a plotline or character that reflected #MeToo themes, but even more provocatively, how these themes were being assimilated into the public face of the entertainment industry. I also wanted to see what the shortcomings were of these efforts to declare their industry as now sensitive to the problem of sexual harassment, but which, much like the "Black Lives Matter" movement, has not been matched by real structural changes in the industry itself to address its racist and sexist underpinnings.

Introduction

The Hashtag That Launched a Movement

When historians look back on the #MeToo movement, the year 2016 will be viewed as an important milestone in its prehistory. In July of 2016, Gretchen Carlson, a former newscaster for Fox News (1996–present) filed a sexual harassment lawsuit against Roger Ailes, who was in charge of the Fox News show. Similar allegations would soon come forward against him, and by July 21 of that same year, Ailes was forced to resign from his position. He was, however, able to publicly deny that he ever engaged in any sexual misconduct and received a severance package well into the several millions. In January of the following year, 2017, another historical milestone occurred on the first day of the presidency of Donald J. Trump, when millions of women protested in Washington for the Women's March in order to support gender equality. This protest was mirrored in cities all over the world and the march in Washington turned out to be what many believed was the largest single-day demonstration in American history. In April of 2017, an article in the *New York Times* came out that outlined five female employees' allegations of sexual harassment against a second Fox News employee, anchor Bill O'Reilly. Once the article was printed, several of the advertisers dropped his show, *The O'Reilly Factor* (1996–2017), and he was eventually fired from his position, even as he claimed that he did nothing wrong and that it was a "political and financial hit job" (Kim 2017).

On October 5, 2017, another critical event in the birth of the #MeToo movement occurred. Jodi Kantor and Megan Twohey (2017) two *New York Times* investigative reporters, published an article titled, "Harvey Weinstein Paid Off Sexual Harassment Accusers for Decades." Weinstein, the founder of the independent film company Miramax three decades earlier, had allegedly been sexually harassing and assaulting women in the Hollywood film industry since the early 1980s. Eventually, over 85 women would come forward with stories that ranged from

3

inappropriate comments and groping to sexual assault and rape. The article revealed that though Weinstein's behavior had been well known for decades, it had been systematically covered up by those who worked for him, and reinforced through non-disclosure agreements which barred the victims from discussing the abuse in exchange for a cash settlement. In response to the article, Weinstein himself went on to publish a public apology, and referenced the fact that in the 1960s and '70s, there were different standards of what was considered to be acceptable behavior in the workplace (Weinstein 2017). Five days after Kantor and Twohey published their account, the reporter Ronan Farrow published an article in *The New Yorker* that also outlined the systematic abuse and repeated incidents of harassment by Weinstein, and cited more than 13 women who had come forward to tell their stories (Farrow 2017). From these articles, which in turn generated widespread media coverage, another phenomenon emerged, the so-called "Weinstein effect," which was understood as a worldwide phenomenon where women who had been sexually abused and harassed came forward to publicly accuse the powerful men in their fields of inappropriate conduct (Graham 2017; Nemzoff 2017). Weinstein was ultimately brought to trial and found guilty and sentenced in March 2020 to 23 years in prison. At the conclusion of the trial, Manhattan District Attorney Cyrus Vance, Jr., remarked on the importance of the women who came forward to testify about the pattern of Weinstein's behavior and said, "We thank the survivors for their remarkable statements today and indescribable courage over the last two years" (Vance 2020). Weinstein, who was 67 when he received the verdict, was described by his lawyers as having "deep remorse," but that he, like other men of his generation, were "totally confused" by events that were brought about by the #MeToo movement (Dwyer 2020).

By October 16, 2017, another event occurred which would signal the real beginning of #MeToo as an identifiable social and political movement—the #MeToo hashtag movement was created on Twitter. It began as a post from the actress Alyssa Milano (2017) which read, "Suggested by a friend. If all the women who have been sexually harassed or assaulted wrote 'Me too' as a status, we might give people a sense of the magnitude of the problem." Milano wanted to encourage other women to speak up about their own stories of sexual harassment. Through making this Twitter post Milano, through no intention of her own, borrowed from another activist named Tarana Burke, who had earlier begun a media campaign on MySpace in 2006 to try to promote solidarity for women of color who had experienced sexual abuse.[1]

In the wake of the Weinstein allegations and the celebrity status of Milano, however, the #MeToo movement transformed into a virtual movement across a wide variety of media platforms. In the next 24 hours, there were over 12 million posts, comments and responses to #MeToo, as reported by Facebook. Within one week, the hashtag #MeToo was used over 1.7 million times in over 85 countries (Jarvey 2017). By the end of the year, the flood of stories continued to pour out from all industries and all corners of the world. While there were earlier attempts to create hashtags that targeted sexual harassment, such as #YesAllWomen and #EverydaySexism, it is noteworthy that these two hashtags generated fewer uses in the whole of 2017 than #MeToo was able to accumulate in its first 24 hours (Main 2017). Different countries had their own hashtags that they used, including the French version, "#BalanceTonPorc" or "rat out your pig." The Arab version, "#AnaKaman," which is roughly translated as "me too," ended up being used by millions of women.

HimToo: Harvey Weinstein, the Hollywood mogul whose sexual assaults of women were so relentless they finally ignited a movement (Wikimedia Image by David Shankbone).

In that first month, the repercussions of the #MeToo movement began to be seen in a variety of industries, in addition to the entertainment and media industries. *Vox* magazine published an infographic that listed over 263 celebrities, politicians and CEOs who were accused of sexual misconduct (Vox 2019). By the end of the year, *Time*'s "Person of the Year" cover featured not a single individual, but rather a group of women who were called "The Silence Breakers: The Voices That Launched a Movement"—referring to the women who spoke up in the wake of the #MeToo movement (Zacharek et al. 2017). In the article, they referred to an earlier moment in American social history, the "problem that has no name," as a precursor to the current #MeToo movement. The feminist author and activist Betty Friedan, more than

50 years earlier, had described the repression and frustration of women of her generation and their lives which were circumscribed by restrictive gender roles as "the problem which has no name." The new problem the writers were now referring to was also not new, in the sense that millions of people had been experiencing sexual harassment in their professional and private lives for decades (Zacharek et al. 2017). And this new movement, much like the women's movement decades earlier, grew out of the recognition of the need to no longer remain silent about their experiences.

In 2018, the impact of the #MeToo movement continued to be felt with Time's Up, an organization that was an offshoot of #MeToo that was formed by 300 women who worked in the entertainment industry. This group announced in an open letter published in the *New York Times* and in the Spanish-language newspaper *La Opinión* that they were going to set up a legal fund to support low-wage women workers who wanted to file sexual assault and harassment cases (Nicolaou and Smith 2019). In January of that same year, another media piece was published which further tried to expose the pervasiveness of sexual harassment in the media industry. A writer named Moira Donegan disclosed that she was the author of a list called *Shitty Media Men*, an anonymous Google spreadsheet that allowed people to detail incidents of sexual misconduct by men who worked in media. At least 70 men who engaged in a wide variety of forms of sexual harassment in the workplace in media organizations were accused in the list. At the same time as these efforts to highlight sexual harassment came forth, Hollywood celebrities were also accused of inappropriate behavior. In the same month, for example, five women accused James Franco of inappropriate behavior.[2] In another #MeToo revelation, an article published on Babe.net detailed a sexual experience by a young female photographer with the comedian Aziz Ansari, which in turn provoked a long public debate about the role of consent in sexual encounters. Ansari responded in a way that would become typical of many men in the entertainment industry, when he explained that he took the "words to heart," and that he "continues to support the movement that is happening in our culture. It is necessary and long overdue" (Nicolaou and Smith 2019).

By the end of January 2018, millions of women participated in the second annual Women's March, and by March of that year, the Oscars ceremony dedicated an entire segment to the #MeToo movement. The Oscars featured Annabella Sciorra, Ashley Judd, and Salma Hayek, who had all accused Weinstein of sexual harassment, and who told the audience they believed the movement needed to bring many changes to Hollywood and beyond. Part of that change was to include more stories by women and people of color, and as Hayek told the audience, "We salute those unstoppable spirits who kicked ass and broke through the bias perception against their gender, race, and ethnicity to tell their stories" (Nicolaou and Smith 2019). By the spring of 2018, the reverberations from the #MeToo movement continued to shake the media and entertainment industries. In April of that year, Jodi Kantor and Megan Twohey ended up winning the Pulitzer Prize for public service for their exposé on Weinstein, as did Ronan Farrow of *The New Yorker*. Women began to respond to the Time's Up Legal Defense Fund, and over 2,500 women contacted them for support for their cases. Entertainers who had been accused of sexual harassment continued to be charged and their shows canceled, including R. Kelly, the male singer who had been accused of running a "sex cult" for decades but whose career had remained intact until 2018, when his concerts were canceled and he was removed from the streaming service Spotify. By June of that year, a crisis consulting firm released a study which found that at least 417 high-profile executives in different fields were accused of sexual harassment. And in August, Ronan Farrow published another article in *The New Yorker*, which described allegations by several women against CBS CEO Les Moonves, who later stepped down. Moonves also announced that he would donate $20 million to the #MeToo movement and organizations that supported it (Nicolaou and Smith 2019).

At the same time that these events occurred, the beginnings of a backlash against #MeToo also began to surface. This could be seen in the case of the comedian Louis C.K., who had earlier been accused of and admitted to having masturbated in front of several female comedians. After a brief hiatus, he went back to doing stand-up comedy while having remained well-loved throughout Europe during the period where he had canceled performances after his sexual behavior had been brought to light. In addition, celebrities such as Kevin Spacey, who had been accused of sexually harassing young men, found that the Los Angeles district attorney's office decided not to file sexual assault charges. Sexual assault charges against actor Anthony Anderson of the television show *Blackish* (ABC, 2014–present) also were dropped by the D.A.'s office. The case against Kevin Spacey was said to be outside the statute of limitations so could not be prosecuted. In the case of Anderson, the woman who had accused him of assaulting her while she had catered an event for him, declined to do any more interviews with the police after filing the initial charge. Anderson, for his part, and who had previously been accused of sexual assault by another woman when he was working on the movie *Hustle and Flow*, denied the allegations (Judge 2018).

Perhaps most revealing of the complex shifts during this period occurred in the wake of the hearings for the Supreme Court nominee Brett Kavanaugh, which included searing testimony from Dr. Christine Blasey Ford, who accused Kavanaugh of sexually assaulting her decades earlier. Ford, a research psychologist from Stanford University, offered testimony describing a harrowing experience of sexual assault at the hands of Kavanaugh that had occurred decades earlier when they were in high school together in Bethesda, Maryland. She said that Kavanaugh, who was very drunk at the time, held her down on a bed at a party, and that as he was trying to undress her, she felt like she was going to be suffocated to death as he tried to cover her mouth to silence her screams. Kavanaugh responded with testimony claiming the assault never took place, in a "boisterous, emphatic denial of any of all claims of impropriety" (Abramson 2018). Despite Ford's testimony against Kavanaugh, as well as that of other women who came forward against him, including Yale classmates Deborah Ramirez and Julie Swetnick, he was nevertheless successfully nominated and became a Supreme Court Justice.

Other signs of a backlash beginning to occur in response to the #MeToo movement were visible that year. Roman Polanski, who had been convicted in the 1970s of sex offenses and who fled to Europe to avoid serving a prison term in the United States, announced that he was going to make a film about a soldier who was falsely accused, based on the Dreyfus affair. The film, *An Officer and a Spy* (2019), was the first film Polanski made after he was eventually expelled by the Academy of Motion Pictures Arts and Sciences under their newly created "ethical standards." Despite his expulsion, the film went on to receive critical praise and multiple awards including the Grand Jury Prize at the 76th Venice International Film Festival on August 30, 2019, while also garnering 12 nominations from the French 45th Cesar Awards (Marshall 2020). Despite winning several awards at the Cesars, Polanski said he didn't want to attend because he was afraid of a "public lynching," and then went on to note that "[w]e know how this evening will unfold already.... What place can there be in such deplorable conditions for a film about the defense of truth, the fight for justice, blind hate and anti–Semitism?" (Marshall 2020). While the backlash against the #MeToo movement included denials by those who had been accused of sexual harassment and assault, it also began to manifest itself in counter-accusations and eventually counter-lawsuits, which arguably had a more chilling effect than simply denying that an assault took place.

In trying to account for the reasons why the #MeToo movement arose and why there has been such a strong cultural reaction, including the counter-response of denials and backlash, it may be helpful to turn to some of the reasons why it took so long, in the eyes of many, to have women come forward to tell their stories of harassment and abuse. Some writers, like Rebecca Traister in her book, *Good and Mad: The Revolutionary Power of Women's Anger* (2018), observed that women who have been silent about the sexual abuse they suffered at different points in their own history were finally granted the cultural space to speak up about the humiliation and pain they had experienced. She noted that the cultural injunctions against women being angry that are rife in popular culture are associated with women being seen, more generally, as hysterical and that "[t]here's a price to pay that people won't take you seriously ... but also that you will pay a price for it." As the popular press and news accounts began to cover these stories of harassment and assault, however, women found that they could finally tell their own stories. The cultural force of #MeToo became part of the larger conversation around women's experience, which was then portrayed not only in news coverage in the popular press, but in scripted and non-scripted television and films as well. More generally, a broad national dialogue is underway regarding how the movement should be defined, what its core purposes should be, and where the parameters of the movement should begin and end. It will take time to sort out the definition and gain some perspective on how the movement may contribute to our sense of gender-based advantages and disadvantages and the direction American society is heading in this regard.

While much attention has been given to the #MeToo movement in the press, relatively less has been written about its impact on storylines and the changing attitudes of creators in the film and television industry. In the wake of recent political, cultural, social and economic trends in the United States and beyond, the changing status of women signals that this as an important area to pursue, particularly in terms of understanding the different genres these storylines and characters appear in, as well as whether more women have been able to create these stories as showrunners, writers, producers and directors. The hope after the #MeToo movement, with its emphasis on exposing the sexism and sexual harassment of people both in front of as well as behinds the screens, was that significant changes would occur in achieving gender parity for women in the film and television industry. The purpose of this book is to examine how core #MeToo movement themes that have achieved some measure of national consensual agreement have been portrayed in popular culture and media as well as how the entertainment industry has responded to the movement itself. Some of these themes revolve around gender dynamics and the abuse of power, of the old sexist culture and the way it manifests itself in relationships between men and women in the workplace and in their personal lives.

The context for this study is the fact that women are still struggling for parity in representation in film and television. In 2018, for example, of the top-grossing films, females comprised only 4 percent of the directors; 15 percent of the writers; 3 percent of the cinematographers; 18 percent of the producers; 18 percent of the executive producers; and 14 percent of the editors. At the same time, they account for 51 percent of the moviegoers (Lauzen 2019a). Women were also underrepresented as protagonists in films, where male characters continued to dominate the screen. What this means in practical terms is that if you were going to see a film in 2018, you were almost twice as likely to see a lead male character as you were a female character in the top grossing films. That is because females only accounted for 36 percent of the major characters in films in 2018, which was actually a one-point decline from 37 percent in 2017 (Lauzen 2019b). Not only are women still underrepresented in top grossing films but when they are shown, they are usually young and white and not in leading roles.

What do these statistics tell us about the impact of the #MeToo movement on film and television? Given the relatively long timeline from writing to production to showing films and television series to audiences, it may be that the dearth of women in 2018 in the entertainment industry means that there might not be many #MeToo stories being told yet. While it may be understandable that the relatively new #MeToo movement hasn't translated more quickly into a flood of films and television shows, it's also possible that film and television executives are still part of a system that minimizes women's experiences and is still resistant to changing the culture to include more women behind and in front of the cameras. They may also be waiting to see what the impact is on the zeitgeist in the larger culture of the #MeToo movement and whether it will be lasting in terms of its impact. For example, Jennifer Kaytin Robinson, the creator of the now canceled series *Sweet/Vicious* (MTV, 2016–2017) recalls pitching a show with a #MeToo theme after the comedian Louis C.K.'s behavior was publicly revealed and highly publicized. She recounts that a male executive said, "This won't be on the air for another 12 to 18 months; do you still think it's going to be relevant? Because it seems like it's getting so much better now." She recounted replying to him, "It is great that we got Harvey Weinstein and Louis, but it's been bad for us since, like, not even the dawn of time.... I do think in 12 to 18 months it will still be bad." In the end, he rejected her pitch (Friedlander 2018).

While the individuals who have been affected by harassment and sexism in the television industry have begun to speak up and been empowered by #MeToo, then, it usually takes some time for film and television programming to catch up with the news cycles and larger world events in their storylines. This is due, in part, to the production constraints around films and scripted TV as well as the fact that writers need to create stories that address sexual harassment and assault. This is true even though the issue of harassment has been ever-present. As David Shore, the creator of *The Good Doctor*, a medical drama on ABC (2017–present) noted, "The issue didn't come up when #MeToo started—the issue's been around, and that's why I think it resonated so much when this started happening" (Friedlander 2018). Despite these constraints around the smaller number of women behind the cameras and leading shows, as well as the lag in production time and doubts about whether this movement will "stick," there have been many stories both in film and television that have explored the mistreatment of women in larger numbers than before the movement. For example, by 2019 and 2020, many films introduced storylines that focused on women's experience with sexual assault or misconduct, from Netflix's mini-series *Unbelievable* (2019) to Apple TV's mini-series *The Morning Show* (2019–present) to Lionsgate's *Bombshell* (2019) to Showtime's mini-series *The Loudest Voice* (2019) to Focus Feature's *Promising Young Woman* (2020) and so on. In addition to film and television dramas, many television comedies also had stories and sketches devoted to issues around #MeToo. Shows like HBO's Season 10 of *Curb Your Enthusiasm* (2000–present) and NBC's 2018 season of *Saturday Night Live* (1975–present) had storylines and sketches devoted to the discussion and Season 8 of *New Girl* (2011–2018) included themes that humorously covered the topic.

In the wake of streaming media companies like Netflix and Amazon Prime, with their tremendous budgets devoted to new media content for their subscribers, there was a pressing need for more stories from all kinds of groups, and women have benefited from this push by seeing their stories appear on screen in numbers seldom seen before. Netflix and Amazon Prime emerged as powerful generators of content about women's experiences in a wide variety of genres. Stories from other films and television series in other countries also reveal that the #MeToo movement, while occurring primarily in the United States, has nevertheless made an impact in other parts of the world and can be viewed, therefore, as having a worldwide influence.

In sum, the purpose of this book is to try to begin to understand the impact of #MeToo on the entertainment industry and the stories it creates. While not an exhaustive analysis of all films and television shows produced in the United States, I focused my lens on exploring those series and films, primarily from 2018 to 2020, which had stories and ideas that drew on the #MeToo movement. My guiding question was how has the entertainment industry tried to portray women's experiences with misogyny and sexism in the wake of the #MeToo Movement? In order to answer that, I looked at specific genres where these stories emerged, including those of drama, documentary, comedy, horror, and reality television and films. Within these genres, I tried to understand how characters and plots drew on the themes of the #MeToo movement and how the storylines embodied these themes. I also looked at how the entertainment industry itself responded to the demands made about addressing sexual harassment and telling more stories about women's lives in the wake of the #MeToo movement. Part of the response included concrete changes in hiring and creating new rules and guidelines that addressed the entrenched culture of sexual harassment that occurred prior to the #MeToo movement. At the same time, there were also concerted efforts to refute and deny the charges launched in the wake of the #MeToo movement against industry insiders, and a backlash did emerge. There were also responses in other countries that both mirrored the reactions in the United States as well as offered their own distinctive cultural responses, and this was explored as well. The relative novelty of the #MeToo movement has meant that there have not been many empirical studies conducted on the ways these narratives have filtered into popular culture more generally. In this way, this study will help to fill in this gap as it looks to the impact of the movement on the film and television industry.

In thinking about how these narratives are being represented in film and television, the theory of media representation, drawn from cultural scholars such as Stuart Hall (1997) are instructive. Hall was an early proponent of the idea that representation, understood as creating meanings through language, is connected to the larger culture. Representations allow us to make "sense" of the world and to understand the actions of those around us and in the larger society. Often expressed through discourse as well as popular media representations, it allows us to derive meaning about society and our societal norms. Another contributor to this understanding of the importance of representations in understanding society is offered by Eugenia Siapera (2010), who understands representation as both the result of media productions as well as something created within the context of the production of media itself. This is especially helpful in understanding the impact of the #MeToo movement in popular culture and society, since the very creators of film and television have themselves been subject to and impacted by its issues.[3]

Nabila Nuraddin (2018), in her study of how international media represented the #MeToo movement, has pointed out that these relationships don't occur within a vacuum, but are rather themselves mediated by who have the authority to speak and who can influence these representations. Cynthia Fuchs (2017) has written about this power dynamic that is inherent in cultural representations, and Manuel Castells (2007) provides a useful analytic framework for thinking about how, within our social structures, there are not only power relations but that these power dynamics are exercised through networks of relations. As we begin to analyze how the #MeToo movement impacted the entertainment industry and the stories that were told both in front of as well as behind the cameras, this understanding of how power works in a post-modern society is illustrative. Fuch's understanding of how those in power in these industries exercise control over the narrative is instructive as well and at the same time, Castell's understanding of how, with the rise of self-mass-communication, there is an opportunity for those who were previously not heard to speak up through these new media platforms, is also helpful to keep in mind. The #MeToo movement, in this sense, is a striking example of both the power of online platforms to get the word out, literally, about sexual harassment, but also allows for more interplay to have these experiences then translated into stories the entertainment industry goes on to produce and disseminate.

In these ways, the discussion of representation and power allows us to understand that the #MeToo movement, by bringing forward the stories of millions of women who had been harassed and assaulted, was able to call into question earlier societal representations of relations between men and women that had either not been previously raised or had been minimized or ignored. At the same time, the #MeToo movement, through bringing these stories out via social media platforms, created a new opportunity to change the power dynamic that has existed in Hollywood and beyond for decades, so that the narratives that would now emerge would have to acknowledge the experience of women from a variety of industries who had been subjected to abuse. These cultural theories of representation and power are useful because they help to explain the ways these stories were translated into popular culture via film and television, and at the same time, served as a "mirror" for changing perceptions about what was considered acceptable and unacceptable forms of contact between men and women in the workplace and beyond.

The chapters in this book try to understand a different way that #MeToo has had an impact on film and television. In the first chapter I look at how #MeToo, which first took root in the entertainment industry, introduced several changes in industry practices in Hollywood. We see how in the wake of the first allegations, there were shake ups that occurred in both the executive suites as well as in the lower ranks of the industry, and how studios, entertainment firms and individuals all tried to respond to the allegations that were coming forward. In the second chapter, I look at how storytelling began to change in the wake of #MeToo. New storylines began to emerge on television shows that referenced directly and indirectly the events happening in real time in the culture around sexual harassment and assault. The question of how far the #MeToo movement has gone, and whether it engaged in a kind of overreach that trampled on the rights of the accused forms the backdrop of Chapter 3. It is shown that, as a response to accusations, many of the key figures who were challenged enlisted the media and the courts to directly counter the allegations against them. More generally, there were voices emerging in the media and entertainment industries that veered from outright denial to a more muted quest for some kind of redemption in the face of accusations that they had behaved in a disrespectful way.

In Chapters 4 through 6, I focus on how different genres drew on themes from the #MeToo era. Genre in this sense is understood as a type or category of film or television show that shares a set of characteristics with other TV programs in that category. There are various theories why these genres survive in different eras, including the need for audiences to have narratives that are stable and predictable as well as innovative. For example, in Chapter 4, I explore the question of how comedy in film and television series drew on humor and irony to address the issues raised by the #MeToo movement. In some ways, as that chapter explores, comedy had more leeway to represent these issues, as it engaged audiences in reflecting on the meaning of these cultural shifts by using humor, whereas reality shows, by contrast, at times themselves became the subject of harassment allegations, given their emphasis on heightening tensions between contestants.

In Chapter 5, I look at how dramas and documentaries unfolded in the era of #MeToo. It was a natural fit to have genres like drama tackle the fallout from #MeToo since the stories themselves created a sense of tragedy and urgency in trying to depict the experience of those who suffered from sexual harassment and assault. Similarly, documentaries also lent themselves to being a primary vehicle for representing the real life events that transpired for different victims of sexual assault. Telling one's truth, in other words, and speaking up about the experience of sexual harassment and assault, was a primary outlet for tapping into an audience's curiosity and desire to learn about people who were in the news and who had been victimized in this way. Chapter 6 explores the genre of horror films and looks at how themes of the #MeToo movement, including the trauma experienced by female characters in these movies, translated into stories and plot devices to explain the action of the characters. At the other end of the spectrum are reality television shows which, while supposedly rooted in the real world instead of the imaginary world of fiction, nevertheless also delivered their own fictional narratives where themes from the #MeToo era could be found.

Chapter 7 moves out into the larger landscape of the impact of #MeToo on international film and television industries by looking at responses by European and Asian industry insiders and audiences. I explore whether there were cultural differences in the kinds of responses to the #MeToo movement that different countries exhibited. In some countries, for example, it was downplayed and viewed as a kind of over reaction by American film and television audiences and industry, while in other countries it resonated and led to new stories being told and a shake-up of earlier assumptions of what is and is not acceptable behavior. The Conclusion looks more broadly at the potential ultimate impact of the #MeToo movement on the entertainment industry and the stories that emerge from it as we move away from these initial responses.

It is in this sense, finally, of trying to understand the cultural shifts that occurred (or failed to occur) in the wake of the #MeToo movement that it may be helpful to also variously refer to it as the "#MeToo era" in this study. This is not because there has not been some kind of a movement that is accurately called #MeToo, but rather that it may also be helpful to describe the period of time after the movement began as the "MeToo" era, to understand how this time has been influenced by the ideas and meanings invoked as part of the #MeToo movement.

1

#MeToo and Its Impact on the Entertainment Industry

In January 2018, a national online survey was conducted around the issue of sexual harassment. One thousand women and one thousand men were asked whether they had been subject to a wide variety of behaviors, including such acts as touching, genital flashing, sexual assault, cyber sexual assault and being followed on the street. The survey found that 81 percent of women and 43 percent of men had experienced one or more forms of sexual harassment at some point in their life (Chira 2018). This figure was higher than had previously been found in earlier studies and polls. One of the differences between this study and other studies was that the survey asked about a larger number of behaviors in a variety of places, from the street to the workplace, in schools as well as online. Another poll conducted around the same time by *The Washington Post*/ABC News reported that more than half of women had been subjected to "unwanted sexual advances" (Chira 2018). This poll, and others like it, indicate that the struggle against sexual harassment that the #MeToo era has come to represent is rooted in the real-world experiences of men and women. It also is arguably a factor in people recognizing that certain behaviors constitute harassment, in a way that may not have been recognized before the #MeToo era. At the same time as the #MeToo movement has brought forward changing norms, there are others who question whether it has made an impact or, on the other hand, has gone too far in its attacks on behaviors that might be considered sexual harassment. Jodi Kantor and Megan Twohey, the two *New York Times* writers who broke the Weinstein scandal, reflected that the movement itself has become controversial and provoked a lot of debate. As they noted, "Basically there are three questions about #MeToo issues that are totally unresolved. One: What kind of behaviors are under scrutiny? Are we talking about Aziz Ansari? Yes or no? Number two: How are we evaluating this information? What are the tools we're using to figure out what actually happened? And number three: What's the punishment or accountability that we're going to use? Each of those three questions is a matter of huge debate" (Landsbaum 2010).

In addition to these unresolved questions, others questions remain about whether the "malaise" in Hollywood that has arisen in the wake of the #MeToo movement is going to usher in real changes, rather than just a sense that Hollywood is temporarily "unmoored" (Barnes 2018). Are those individuals who wield power in the industry just biding their time before returning to the old way of doing things, from the proverbial "casting couch" culture to the blocked opportunities women have experienced there for decades? After the Weinstein scandal broke, there was a series of firings and resignations of people, including not only Weinstein himself, but actors and studio heads such as John Lasseter, directors such as Brett Ratner and chief executives including Leslie Moonves. While people in Hollywood spoke approvingly of these people leaving the entertainment industry, privately there was a palpable sense of anxiety and resentment about the ways in which people's careers were upended in the face of accusations of sexual harassment and assault. As one leading film producer, speaking anonymously to a *New York Times* reporter said, "Yap, yap—go back to your kennels," referring to the organization Time's Up, founded by Shonda Rhimes and Reese Witherspoon and other prominent women in Hollywood to fight workplace harassment (Barnes 2018). In these ways, while some people are lauding the changes and others are reacting with a sense of unease and anxiety, it is not clear yet whether these actions will lead to lasting changes in the industry.

It is helpful, though, as we weigh whether the changes will be permanent or not, to remember the original roots of this struggle and the work that had been done historically in fighting sexual abuse, including events like Anita Hill's testimony against Clarence Thomas in the 1990s or Paula Jones' testimony against then President Bill Clinton, also in the 1990s (Kaminsky 2019). This is not to say that the fight against sexual harassment only came about in the wake of #MeToo. The struggle against sexual abuse and sexual harassment first began in the United States in the early 1970s, when laws were put into place that made sexual harassment in the workplace illegal. Despite the fact that there was breakthrough legislation over 40 years ago, it was arguably not until the #MeToo movement that sexual harassment laws finally came into the mainstream as a powerful force for combatting sexual harassment. As Catherine MacKinnon, the legal scholar who was responsible for much of this legislation, noted, "the [#MeToo] movement is surpassing the law in changing norms and providing relief that the law did not ... sexual harassment law prepared the ground, but #MeToo, Time's Up, and similar mobilizations around the world—including #NiUnaMenos in Argentina, #BalanceTonPorc in France, #TheFirstTimeIGotHarassed in Egypt, #WithYou in Japan, and #PremeiroAssedio in Brazil among them—are shifting gender hierarchy's tectonic plates" (MacKinnon 2019).

In other words, while the laws have been on the books for decades, the #MeToo movement has been more effective in "changing norms and providing relief" in a way that was not possible before the #MeToo movement came into being. The beginnings of the #MeToo movement was paved through these women's individual efforts, which created the legal space to move forward with charges of sexual harassment. The problem was that, before this movement, women still had to face a social world where their harassers were more often believed than they were, and where they often times faced retaliation. This is simply to point out that, from the beginning, as important and socially transforming as the #MeToo movement has been, it is an ongoing and uphill battle to eradicate sexual harassment all at once. However, the fact that women's voices are being heard and their accounts of sexual abuse are finally being believed, does signal that a profound shift in the larger culture has already occurred.

In many ways the beginning of the #MeToo era occurred in the film and television industry itself, with the breaking of the story about Harvey Weinstein, one of the most powerful men in Hollywood, and the Hollywood actresses who came forward to speak about his sexual assault and harassment.[1] Perhaps more tellingly, as Weinstein tried to defend himself, he found that his network of industry friends and people he had supported politically would not come to his defense. The industry itself soon found that it was in the midst of tremendous upheaval as allegations began to pour in about people who were in powerful positions and who had committed sexual harassment and assault against other people in the industry. The list of people accused and who eventually left or were fired included such diverse men as Oscar-winning actor Kevin Spacey; political writer Mark Halperin, a former *Time* magazine employee; top executive Brett Ratner from Warner Bros.; Roman Polanski and Bill Cosby, who were both removed from the Academy of Motion Pictures. Les Moonves, the head of CBS, was also investigated and forced to leave.

In the wake of these initial entertainment industry firings, there were also larger repercussions in other industries, including media and the tech industries, among others. Women began to file harassment complaints, and also came forward with their own #MeToo stories.[2] In addition, there was a rise in internal complaints filed with human resource departments of various companies. As a response to the pervasive culture of blocked opportunities and harassment and inequality, the #MeToo movement and the female entertainers impacted by the movement, created another group called Time's Up, an organization devoted to ending sexual abuse in the film industry. Times Up has helped to focus the spotlight on the need for a safe and equal workplace, and along with the National Women's Law Center, also created a defense fund for low-income women to receive attorney consultations and to help pay legal fees involved in filing sexual harassment suits against employers. The fund raised over $22 million in 2018 and was able to help over 3,500 women across the country (Chiwaya 2018).[3] They also donated money to other women's organizations and brought some of these women activists as "dates" to the 2018 Golden Globes to further highlight the work done by these organizations. Another part of the work of Time's Up was to help the Producers' Guild of America draw up anti–sexual harassment guidelines, which have also been created in the UK and in Europe with unions associated with the film and television industries there. Part of the work of these guidelines is to include anti-harassment training for the cast and crew of productions, as well as to identify people to whom workers can report any incidents. Additional groups, such as the Screen Actors Guild, have also included new policies and procedures to deal with harassment (Hutchinson 2018).

Though individual actors in the entertainment industry may not be permanently ousted from Hollywood, it may be more helpful to think about whether #MeToo has resulted in any substantive changes in terms of the laws that impact the behavior of individuals working in the entertainment industry itself. Since the #MeToo movement began, there have been changes in legislation with the intent of offering more protection against sexual harassment. This is especially critical because the entertainment industry, unlike some other fields, has more ambiguity in terms of what constitutes traditional workplace hours or places where the work is conducted. Since work is often performed outside of a traditional 9–5 office space, people who work in entertainment need to be even more conscious of what might be considered inappropriate and illegal behavior. To this end, there are new legal requirements that have been put in place. In New York, for example, where Title VII of the Civil Rights Act of 1964 had already made sexual harassment illegal, there are now new protections which have been put in place, including a requirement for an anti–sexual harassment policy and training program for all their employees, no matter what the industry (Page 2019). In addition, if a freelancer or consultant is sexually harassed, the employer can also be held liable, if it was found that the company was aware or should have been aware of the harassment and "failed to take immediate and appropriate corrective actions" (Page 2019).

In California, similarly, the Fair Employment and Housing Act (FEHA), which applies to employers who have five or more employees, also provides protections against harassment. California already had harassment training in place before 2018, but these laws have now been expanded and require all companies with five or more employees to offer harassment training, whether they are supervisors or hold other positions. There are now additional safeguards to the original laws, which include such features as finding that a single instance of harassment can be the basis of a claim of a hostile workplace environment, as well as that individuals can be held personally liable and that the employers can also be held liable for the actions of nonemployees if the employer should have known or did know of the conduct of the harasser (Page 2019). These kinds of new rules are especially helpful in an industry like the entertainment field where the workplace is often ill defined and where a party can still be viewed as a "workplace" site. By lowering the legal standards of what is considered a hostile workplace, such behavior, which might be engaged in at a film festival or a cast party, is now considered as potentially liable as if you were at a traditional office.

Morality Clauses

Morality clauses are another legal strategy being used by film companies in the wake of the #MeToo movement. There is now language included in contracts with stars, filmmakers and distributors that give the company the right to drop individuals if they are accused of misconduct. The term used in the contracts is "moral turpitude," defined as "an act or behavior that gravely violates the sentiment or accepted standard of the community" (Siegel 2018). Several movie studios, including Fox and Paramount, are now putting morality clauses into their talent deals, which would allow them to terminate a contract if "the talent engages in conduct that results in adverse publicity or notoriety or risks bringing the talent into public disrepute, contempt, scandal or ridicule" (Siegel 2018). In addition to these bigger film companies, smaller distributors are also adding these kinds of clauses, and studios and buyers are all trying to deal with the fallout of financial losses that have been incurred in the wake of scandals when individuals were accused of misconduct. These include losses that resulted in the wake of accusations against Kevin Spacey and Louis C.K., Jeremy Piven and Brett Rattner, among others. For lawyers who are trying to protect their distributor clients, there is an impetus to include these morality clauses into contracts in order to protect them if someone involved with the film is accused of sexual harassment. For those lawyers who represent the talent, however, there is resistance to these clauses because if someone is accused, then the company can fire them and not pay them. Some lawyers also have found that actors and directors can also suffer if the studios and buyers they work for are similarly accused of harassment, as was the case with Harvey Weinstein. They object to this seemingly double standard—their clients have to be bound by these clauses but the companies their clients work for don't have the same requirement of their executives (Siegel 2018).

Clearly, without morality clauses, companies can suffer financially if there are sexual harassment allegations leveled against the talent they hire, even if they go ahead and fire them anyway. For example, after Kevin Spacey was accused of assault by numerous people, Netflix decided to fire him from *House of Cards* (2013–2018), but he ended up being paid for the entire season of the show, even though he did not appear in the final season. Netflix took a $39 million loss in letting him go. In addition, Spacey originally starred in the movie *All the Money in the World* (2017), but when it came to light that he was accused of harassment, the company that financed the film, Imperative Entertainment, decided to pay $10 million to replace him with Christopher Plummer and reshoot the film (Siegel 2018). More generally, distributors are losing money when #MeToo harassment claims are leveled against the talent they hired, such as the case of what happened when *The Birth of the Nation*'s (2016) star and filmmaker Nate Parker was later found to have been accused and tried for rape when he was in college. Although he was acquitted of the charges at the time, the accuser subsequently took her life, and when these revelations came out, Fox Searchlight, who distributed the film after it was first shown at the Sundance Film Festival, ended up losing millions of dollars when they decided not to release the film. Other films have also suffered in the wake of allegations, as when Louis C.K.'s *I Love You, Daddy* (2017), which had been acquired by the Orchard for $5 million, was not released in the wake of harassment allegations against him. Louis C.K. ended up paying back the Orchard for what it had originally put up, even though he was not legally obliged to, since there wasn't a morality clause in his contract.

Non-Disclosure Agreements

One of the most important developments that has occurred in terms of enacting structural changes in laws in the wake of the #MeToo movement revolves around the widespread use of non-disclosure agreements in settlements where sexual harassment was found to have taken place. Non-disclosure agreements, also referred to as "NDAs," or confidential provisions, are part of settlement agreements in business that are a legally enforceable way to keep silent about something between the two parties (Khorasani 2018). NDAs victims signatures effectively silenced them about their experiences of sexual harassment and assault. The *New York Times* revealed on October 5, 2017, that Harvey Weinstein had non-disclosure agreements with eight women, dating back to 1990 when the victims were paid money and were in turn prohibited from speaking up about what had occurred. Because these agreements were signed, the public was unaware of these acts, and they continued to occur until 2017 when Weinstein's behavior was exposed.

What the Weinstein story, as well as the many stories that came after it, also revealed was that NDAs were not limited to one producer, but rather occurred throughout the entertainment industry—other high-powered individuals had NDAs in place to protect them from their victims speaking out as well. The silence of these voices occurred as part of a larger Hollywood culture and was reinforced by the public perception that the entertainment industry was a glamorous industry where powerful people were able to make underlings. People wanting to "make it" in Hollywood were therefore put in a position to accept inappropriate behavior in order to obtain work. The "casting couch" is a concept that is also part of popular culture and is seen as part of the way powerful people in the entertainment industry can exploit people to obtain sexual favors for a job. Because of this culture, the burden was often placed on the victims to deal with the fallout of the abuse, or as Mira Sorvino, an actress who herself was subject to Weinstein's harassment, put it, "the entire system always put it on the young person entering it … to navigate the shoals of sexual harassment alone … this is your problem, you need to deal with it. Don't make too much noise, just deal with it" (Khorasani 2018). This led to many of the victims never reporting it in the first place for fear of retaliation, or when they did, finding that they were pressured into settling and signing non-disclosure agreements.

One of the most negative effects of these NDAs, then, was that they allowed perpetrators to not be held accountable and to continue their behavior with new victims. For example, in one of these cases Weinstein abused a young woman, Zelda Perkins, when she worked at Miramax almost two decades before the *New York Times* article came out. At that time, Weinstein asked her for a massage in his hotel room while he was dressed in only his underwear, and when she refused, he then asked her to be in the same room with him while he took a bath. This scenario was a common ruse on Weinstein's part, but she didn't know this at the time and didn't seek legal help until she heard that Weinstein attempted to rape another woman in 1998. Though she wanted to go to Disney, the parent company, her own lawyers discouraged her from doing so, and instead told her that her best strategy would be to get a monetary settlement from him (Prasad 2018). By agreeing to the settlement, Perkins had to sign a NDA, which while legal, made Perkins feel, as she later disclosed in an interview, that it was immoral and that she was isolated in the face of the legal power of Weinstein's lawyers (Prasad 2018). This led her, over 19 years later, to break the terms of the NDA and speak up about what happened while she worked at Miramax. After the allegations against Weinstein were made public in 2017, Weinstein and the Weinstein Company's business practices became the object of investigation, as the settlement agreements and NDAs they had made through the years were made not just by Weinstein himself, but by the management of the company as well as the board (Khorasani 2018).

Given this culture of silence and the prevalence of NDAs, one of the most positive effects of the #MeToo movement has been some shifts in employment laws and IRS rules that in some states have made it illegal to pay individuals for silence about harassment claims. There are laws being created in several states that specifically target laws that gag complainants (Guthrie 2019). In California in 2018, then Governor Jerry Brown signed a law called STAND or Stand Together Against Non-Disclosure Act that bans "confidentiality clauses in settlement agreements that involve sexual assault, sexual harassment or sex discrimination" (Guthrie 2019). New York also enacted laws in 2018 that expand anti–sexual harassment protections. Speaking to the impact of #MeToo on these changes in the law, Ann Fromholz, a California employment attorney offered that "#MeToo has really changed the playing field … before the #MeToo movement and the backlash against these settlement agreements, it was almost universal that confidentiality agreements covered the terms of the agreement but also the facts of the complaint" (Guthrie 2019).

Another way people responded to the NDAs in the wake of #MeToo was to ask to be released from the original non-disclosure agreements that they had signed. For example, NBC Universal agreed in the fall of 2019 to let their former staffers get out of their original non-disclosure agreements (Haring 2019). After Rachel Maddow announced on her MSNBC show (*The Rachel Maddow Show*, 2008–present) that her parent company was allowing people to get out of these agreements, former staffers from Fox News (1996–present) also asked that they be released from their earlier agreements. Gretchen Carlson, who was portrayed in the film *Bombshell* (2019) and who was the first of the women at Fox News to sue Roger Ailes, was quoted as saying, "All women at Fox News and beyond forced to sign NDAs should be released from them immediately, giving them back the voices they deserve ... none of us asked to get into a workplace dispute. We simply had the courage to stand up and say something—but in the end, it's our voices no one can hear. Because of our NDAs, we can never say what is factually correct or incorrect about what happened to us at Fox" (Haring 2019). NBC Universal itself only decided to allow former NBC News employees to be released from their confidentiality agreements after Ronan Farrow published his book *Catch and Kill* (2019), which included charges about improper sexual misconduct that occurred at NBC News, including an account about Matt Lauer, the former *Today* (1952–present) show host, who was accused of forcibly raping a producer. Several other women at Fox News who left and who had signed NDAs also wanted to be able to speak out, including Julie Roginsky and Tamara Holder (Haring 2019). Despite their request, Fox News has not committed to lifting the NDAs the women signed.

Intimacy Coordinators

The entertainment industry, as an employer, is trying to think about ways to create a safe workplace environment not only on the set but also in the scenes themselves in which actors have to play their parts. This includes sex scenes as well as scenes where actors and actresses perform nude or semi-nude. For example, in January of 2018, Emily Meade, who costars with James Franco on the HBO's series *The Deuce* (2017–2019), a television series that focuses on the sex industry in the 1970s, was trying to figure out what to do about continuing to work with Franco because he had been accused of sexual misconduct. Specifically, there were allegations that he had pressured students in his acting classes to perform nude and semi-nude (Bradley 2019a). Franco defended himself by saying the accusations weren't accurate, and after reviewing the allegations, the producers and HBO decided to go forward with the series (Bradley 2019a). Despite the fact that HBO said they were "comfortable" continuing with Franco as the star, Meade asked if HBO could create a position for someone on the crew to specifically focus on and help orchestrate the sex scenes in the series. HBO responded by bringing in an "intimacy coordinator" and implemented a policy that requires an intimacy coach to help coordinate any programming that includes adult content on their shows (Bradley 2019a).

In the wake of this action, other networks and streaming platforms also created similar policies that would help coordinate the scenes that included sexual content on their shows, and SAG-AFTRA is also working on creating guidelines for all their guild productions in the future. SAG-AFTRA president Gabrielle Carteris said that their organization had been considering this issue before #MeToo, and before the rape allegations against Harvey Weinstein. Before that time, intimacy coordinators were usually employed for stage productions rather than for television series. In part, the time lag between stage and television was a result of the fact that television series had earlier not contained much nudity or sexual content, or as she noted, "Look at television now, versus 20 years ago, I mean it's so sexually explicit now.... Think about before I began [acting]; we had to sleep in separate beds, and you couldn't show a woman being pregnant. Now it's everything" (Bradley 2019a). This includes things like kissing, where before it was clear where you could and couldn't touch someone while you were kissing them, but now the pressure is to make it look as sexual as possible.

Ita O'Brien, who serves as an intimacy coordinator for Netflix's series *Sex Education* (2019–present) said that some of the things that they work on in terms of guidelines include making sure that the directors and actors have a conversation about all of the intimate scenes before they sign a contract. In addition, they continue to discuss it during rehearsals as well as performances. Another requirement is that the sex only occurs on a closed set (without the presence of anyone who is not crucial to filming the scene) when the scenes are being filmed. They also make sure to have a conversation with the actors about what makes them uncomfortable, as well as knowing what the nudity clauses are in each actor's contract. Finally, the intimacy coordinator also talks to the director to understand why the scene exists in the first place and whether it is necessary as part of the arc of the character's development (Bradley 2019a). While there is a focus on the actual physical act of sex scenes, part of the work of the intimacy coordinator is to also assess whether anyone is becoming uncomfortable during a scene. Checking in with the actors and trying to help them understand what they do and don't feel comfortable with is a relatively new idea in the entertainment industry, where actors were previously not in a position to speak up about whether or not sexual content made them uncomfortable. And, while there may have been some initial concern about whether this might intrude on the artistic freedom of the directors, Meade herself argued that the safety of the actors is also important, in much the same way as the use of stunt doubles occurs on the set if it thought to be necessary. Stunt doubles, by contrast, have never been viewed as curbing the freedom of the director.

Changing the Scene from Inside and Out

One of the primary ways that the #MeToo era can be felt is through the push to have more women behind the scenes in Hollywood. This includes mandates from networks to have women be directors, as well as a call for more women to be producers. Studios are also running mentoring programs, and actresses are also pressing for a more prominent role in producing shows (Richwine 2018). This is due to the fact that, before this time, there was a dearth of women who were in position to help shape what kinds of stories would get told in Hollywood. There is a sense that there is more receptivity to having women play these roles in the wake of the #MeToo era. Specific examples of this include Susanne Daniels, the global head of original programming at Alphabet Inc.'s YouTube. She has said that she now requires the platform's original series to hire female directors each season (Richwine 2018). Another initiative includes Comcast Corporation's NBC, who created an initiative called "Female Forward" that lets women learn from a director of an NBC show and allows them to direct one episode (Richwine 2018). Another possible structural change that was called for in the wake of the #MeToo movement but which has yet to be adopted was one suggested by the actress Frances McDormand in her 2018 acceptance speech for her Oscar for best actress. This is the so-called "inclusion rider," which was first developed by Stacy Smith of the Annenberg Inclusion Initiative of the University of Southern California in 2018. In her speech McDormand suggested that there should be a contractual obligation that actors and actresses could include to increase the amount of women and minorities in film productions. In accepting her Oscar, McDormand asked for it explicitly, "You can ask for or demand at least 50% diversity in not only the casting and the crew.... I just learned that after 35 years in the film business—we aren't going back" (Hutchinson 2018).

Unfortunately, the "inclusion rider" still has not been used in more than just a few productions (Buckley 2019). There was one independent film, *Hala* (2019), made by Minhal Baig and acquired by Apple at the Sundance Festival, that used the rider as well as another female-led company called Level Forward, which produced a feature called *American Women* (2019). While film companies like WarnerMedia have said that they have a new inclusive hiring policy for films like Michael Jordan's *Just Mercy* (2020), they still don't have an actual rider. For the Netflix series that Jordan has done, *Raising Dion* (2019–present), there is a similar idea of an inclusion approach but without the signing of a specific rider. Even Frances McDormand herself, who through a representative said that she will put them into any project that she will produce, has still not put in the rider. There are several reasons why it may not be to a studio's advantage to have a rider, including the fact that it implies that there are going to be quotas and that if they don't fulfill them, they would have to pay financial penalties that would go into a scholarship fund. In addition, Buckley (2019), citing Paul Feig, the director of *Bridesmaids* (2011), noted that there is the idea that "people in the industry get nervous about anything that limits their ability to have the freedom of choice to hire who they want to hire." While riders are not being signed, there are nevertheless many films that are being directed now by black directors and also have women in starring roles. This is due, in part, to diversity hiring plans that are now in place for behind-the-scenes jobs in companies like WarnerMedia, as well as Paramount and Endeavor Content.

There have also been several showrunners, such as Shonda Rhimes and Ava DuVernay, who hired female directors, as well as Ryan Murphy, the creator of FX's *American Horror Story* (2011–present) who also have created initiatives to hire more women and LGBTQ people and people of color for work behind the scenes. At the same time, these companies are not using riders and justifying it because they say they have put inclusion policies in place, but if they are not being required to then it is ultimately a voluntary effort that leaves companies accountable only to themselves. Another strategy, finally, is to utilize something called a Reframe Stamp, which is a measure that awards points for film and television productions that hire women for on- and off-screen jobs. Created by the group called Women in Film Los Angeles and the Sundance Institute, and begun with the industry site IMDbPro, the system has awarded such films as *Girls Trip* (2017), *Wonder Woman* (2017), and *The Shape of Water* (2017) for their inclusive hiring policies. Since 2018, it has awarded over four dozen more films, including *Crazy Rich Asians* (2018), *Bumblebee* (2018), and *I Feel Pretty* (2018), and the hope is that audiences will be looking for the stamp when they go see a film (Buckley 2019).

There are similar efforts at diversity in terms of gender in television show pilots in 2018, where, for example, at least 14 of the 42 drama show pilots ordered in the spring were directed by women. In the prior year, there had been only one show. These figures suggest that in the #MeToo era there may be more opportunities for women being created, and there is the sense that more doors are opening and female-driven projects are being "green-lighted." Women writers are being hired, and female-driven shows are being piloted, including returns of earlier female-led shows such as *Murphy Brown* (CBS, 1988–1998) and *Cagney*

& Lacey (CBS, 1982–1988). While not all of these pilots will be sub-sequently picked up as television series, there is at least the opportunity for networks and advertisers to potentially choose them over other shows. In 2018, at the Sundance Film Festival in Utah, another positive sign was that 37 percent of the feature films shown were directed by women, and though many of these films will not make it through to distribution to theaters, there is at least the chance that some of them will (Morris 2018).

Though there have been positive changes in these areas, other data suggests that the amount and pace of change is still slow. According to the Center for the Study of Women in Television and Film at San Diego State University, in the 2016–17 TV season the percentage of women who had behind-the-scenes roles as directors, writers, producers and creators was just 28 percent. This is the same figure as when a similar study was done four years earlier. Despite these relatively low figures, the impact of the #MeToo era on the entertainment industry has been arguably greater than on any other industry. This is due, in part, because of the initial reports of harassment at the hands of the Harvey Weinstein, but also because of the ways in which the "casting couch" has been part of the culture of the entertainment industry almost from its inception. For many, the way to address this toxic culture is to hire more women as writers and directors and crew members. Leah Meyerhoff, a founder of a support group for female directors called Films Fatales, noted that "clearly when the majority of people in power are these able-bodied white men, a side effect is sexual harassment on set and in the world" (Morris 2018). One of the big fears among those in the industry is that there is so much more to do and such a long way to go to create meaningful change. The culture is changing, but the pace of it may not be as quick as is hoped. Speaking to the fact that some changes have occurred, though, the actress Dame Helen Mirren, for example, told the BBC that in her early career, while the vast majority of people were men who were on the set, that has lowered to about 75 percent, and that while this is still an uneven number, "there is definitely a change coming without question" (Morris 2018).

Changes in Hollywood Culture: Two Steps Forward, Three Steps Back?

One way to look at shifts in the entertainment industry, ultimately, is to focus on the ways in which it is trying to mitigate and improve on its practices, both informally and formally, to move away from some earlier practices that tolerated and even encouraged sexual harassment. There has been, more generally, a push to have greater diversity in casting, and this is also due to the #OscarsSoWhite controversy, where the issue of having the vast majority of nominations and awards go to white actors and industry professionals led to a public outcry over the pervasive discrimination against people of color in the entertainment industry. Other changes have included rules around where auditions should take place; for example, no longer having them in hotel rooms. Still other industry changes focus on responding to offensive tweets and jokes on Twitter. One example was the firing of new *Saturday Night Live* (NBC, 1975–present) actor Shane Gillis, when it was discovered that he had tweeted and commented in a negative way about women and Asian minorities a year earlier. As the *SNL* spokesperson commented on behalf of Lorne Michaels, the creator and producer of *SNL*, "we want 'SNL' to have a variety of voices and points of view within the show, and we hired Shane on the strength of his talent as comedian and his impressive audition for 'SNL.' We were not aware of his prior remarks that have surfaced over the past few days. The language he used is offensive, hurtful and unacceptable. We are sorry that we did not see these clips earlier, and that our vetting process was not up to our standard" (Gonzalez and Friedlander 2019).

This immediate firing upon the discovery of offensive tweets about women and minorities contrasts sharply with an earlier instance when comedian Trevor Noah was able to replace Jon Stewart even after offensive tweets Noah had written earlier surfaced. These included such tweets as on February 5, 2015: "'Oh yeah the weekend. People are gonna get drunk & think that I'm sexy!'—fat chicks everywhere," or another on October 14, 2011: "A hot white woman with ass is like a unicorn. Even if you do see one, you'll probably never get to ride it" (Diaz 2009). Not only are entertainers being fired for offensive comments, but directors as well, including James Gunn, who Disney fired in July 2018 from the *Guardians of the Galaxy* series when it was discovered that he made offensive jokes on Twitter (Barnes 2018). On the legal level, there are efforts being made by studios and media companies to gain more legal protection if they are charged with hiring individuals who commit sexual harassment. At the same time, however, there are others who question whether the money that these actors generate for Hollywood will in time create a counter-incentive or regression toward earlier bad behavior. And, even without the loss of revenue, there are signs that the movement has not been able to overturn the power structures that still dominate Hollywood. Early signs show that there are, in fact, fewer women running movie studios than before the #MeToo era, as for example, Stacey Snider, the chief executive of 20th Century–Fox who was replaced when Disney took over. In addition, in the year since the #MeToo movement began in 2017, nine more female senior executives left their positions, including such top-ranked women as Diane Nelson, who was president of DC Entertainment, and Cyma Zarghami, who was president of the Nickelodeon Group (Barnes 2018).

Celebrity responses in the #MeToo era have been central to arguing for changes in the entertainment industry. They have been there since the beginning, when the first famous actresses came forward to speak against Harvey Weinstein. Since that time, there has been an outpouring of responses from women and men in front of and behind the screens, arguing for more fundamental changes to occur in industry practices that have been in place for decades. For example, on January 17, 2018, Oprah Winfrey gave an acceptance speech at the Golden Globes ceremony. In the audience were celebrities who had all chosen to wear black outfits as a show of solidarity and a symbol of protest against the sexual harassment and assault women have suffered. As the men and women listened to her speech, she spoke about the ways that the #MeToo movement had been able to call attention to sexual harassment and made a call to action to people from around the world to end sexual harassment and assault. Before this speech, an earlier public demonstration by celebrities included a Times Up manifesto signed by women in the entertainment industry that put forward a denunciation of the pervasive sexual harassment and assault experienced by women and the need to push for enforcement in laws against harassment and gender discrimination in the workplace.

Oprah began the speech by offering an autobiographical description of her own upbringing, which was impoverished and characterized by racial and gender discrimination. She noted that seeing Sidney Poitier receive an Oscar in 1964 was a transformative experience for her, since it was perhaps the first time that she saw a person of color be recognized in that way. She then noted that while Sidney Poitier was able to receive the Cecil B. DeMille Award for lifetime achievement at the Golden Globes in 1982, it was a full 36 years later that the first black woman, herself, would be given the same award. She then discussed the ways that black women had been subjected to sexual violence throughout the history of the United States, citing in particular the case of Recy Taylor, a black woman who, in 1944, had been raped by six white men, but the men were never charged for their crimes. In addition to Taylor, Oprah also spoke about Rosa Parks, the civil rights activist who fought to end segregation and was herself sent to investigate the crimes committed against Taylor. By speaking about the experience of Taylor and the ways that Parks fought for Taylor and other black women, she then pivoted to the current situation affecting women, noting that speaking up about one's experience and not being believed or having it be denied is what links these women in history to the current moment. As she observed, "It's one [story] that transcends any culture, geography, race, religion, politics, or workplace. So I want tonight to express gratitude to all the women who have endured years of abuse and assault because they, like my mother, had children to feed and bills to pay and dreams to pursue. They're the women whose names we'll never know ... they are working in factories and they work in restaurants ... they're our athletes in the Olympics and they're our soldiers in the military" (Winfrey 2018).

Oprah then discussed the #MeToo movement and the Time's Up campaign, and noted that the original term "Me Too" came from a black female activist named Tarana Burke in 2006, but was adopted by mostly white women in the entertainment industry to bring attention to the ways they had been harassed and assaulted. This referencing of the co-opting of the term "#MeToo" by white female entertainers echoes, in turn, how other women's groups also felt that attention had only been paid to their plight once actresses came forward with their stories. Time's Up acknowledged, in turn, that the #MeToo movement, which was led by women in the entertainment industry, needed to incorporate more women in other industries. At the same time, Oprah's speech also acknowledged the ways that her special status as a celebrity allowed her to push for changes in the entertainment and other industries where sexual harassment occurs. Her influence throughout her career, and "The Oprah Effect," her ability to influence people's choices in everything from selling books to helping presidential candidates like Barack Obama, has arguably given her an even greater pulpit from which to advocate for change. This belief in the power to effect social change was clear in her concluding invitation to those in the viewing audience when she offered, "So I want all the girls watching here, now, to know that a new day is on the horizon! And when that new day finally dawns, it will be because of a lot of magnificent women, many of whom are right here in this room tonight, and some pretty phenomenal men, fighting hard to make sure that they become the leaders who take us to the time when nobody ever has to say 'Me too' again" (Winfrey 2018).

In the end, however, in terms of the culture of the entertainment industry, it is hard to find the same level of attention and support that was given during the big push at the Golden Globes ceremony. Individuals who have been identified as engaging in sexual misconduct have faced repercussions, such as director Bryan Singer being fired from *Bohemian Rhapsody* (2018), or Woody Allen having *A Rainy Day in*

New York (2019) not scheduled by Amazon, which had funded its production. At the same time, other careers have been left intact, and the number of women in key industry positions of power and authority has reversed. One of the ways that Hollywood has responded, as will be shown, is through interweaving stories of sexual harassment and assault as plot devices in the narratives they present on screen. The question remains, however, whether these shifts in storylines will be mirrored in any meaningful way in fundamental changes in the industry itself. As Amy Baer, the president of Women in Film, noted, "Established protocols—decades worth—are changing at lightning speed … for people like me, who believe change is desperately needed in Hollywood, that is exciting. But a lot of people are lost in anxiety" (Barnes 2018). Still, there may be hope that there are seeds of change which are being planted and which will continue to grow, as the movement matures and as new generations of film makers and creators enter the field.

As one indication of this, it is instructive to note that many of the top film schools in the years following #MeToo's first revelations are now addressing the issue directly in their classes, where they hold special discussions with faculty, students and alumni. As Elizabeth Daley, who serves as dean of the School of Cinematic Arts at the University of Southern California noted, "if we want a different kind of industry, we have to be able to envision it. You can't create what you can't imagine, so I think the important thing for us is not only to acknowledge the problems, but also to talk about what kind of industry we want. Where do we want it to go? How should we work to treat one another so that we can realize that industry?" (Prange 2018). Other deans have echoed similar sentiments, including Teri Schwartz, dean of the UCLA School of Theater, Film and Television, as well as Allyson Green, dean of the NYU Tisch School of the Arts. In addition to sponsoring programs that focus on diversity and inclusion, these schools are also holding panels that include diverse professionals and women telling their stories. They are also, not surprisingly, turning down money from people like Weinstein, who had made a $5 million pledge to USC to fund an endowment that was supposed to go to women filmmakers. In addition, USC also accepted a request from director Bryan Singer to not use his name for the Division of Cinema and Media studies, as he was facing his own allegations of sexual misconduct (Prange 2018). On a more fundamental level, these schools are trying to encourage women filmmakers to tell their own stories, and there are a large number of women film students relative to the 1960s. The deans of these schools, who are all women, were able to identify with the goals of the #MeToo movement, since they had all worked in the film industry for generations, and were able to speak to the reality that they wouldn't have been believed in earlier eras if they had come forward with harassment claims. As Green observed, "It has to be talked about. It has to be identified. It has to be named. It has to be called out" (Prange 2018). In these ways, it is perhaps possible to imagine a different entertainment industry, one where the younger generations are not afraid to speak up and to call out harassment where they see it. Whether this translates into a wholesale shift in industry practices, however, remains an open question for now.

To summarize, the structural problems which have plagued Hollywood for decades, including the lack of women in executive roles as well as creative areas, remain. Though this may take a long time to overturn, there have been changes in industry practices, as well as an awakening about earlier behaviors toward women. This has occurred despite the fact that many of the men who believed they were wrongly accused charged that the #MeToo movement had gone too far. Noting that real change has occurred in Hollywood, Melissa Silverstein commented that "[t]he bottom line is, the tectonic plates of the industry have shifted completely ... there is always going to be this understanding that egregious things have happened across multiple parts of this industry and people can see that" (Faughnder and Perman 2020). Some of these changes have included new guidelines in the agencies and guilds such as SAG-AFTRA that have been created to protect their members from sexual harassment. On the industry side, there are also attempts by studios to be more careful of who they hire, and they are including morality clauses in their contracts with them. The Academy of Motion Picture Arts and Sciences have also instituted new standards of conduct and ejected members who had been convicted of harassment. On the legal end, there are now new laws restricting the use of non-disclosure agreements and several states have expanded the statute of limitations for sex crimes, including New York and California. And the conviction of Harvey Weinstein has signaled that the culture of the "casting couch" is no longer accepted in Hollywood as an excuse for sexual assault. At the same time, the entertainment companies have also fired several individuals who have a history of sexual harassment and have hired women for positions of power in the wake of these firings. Despite these advances, other women have left important posts, and the consensus is that for real change to occur, putting more women in power and having them share power with men will be critical to enacting true reforms.

In terms of social activism, we have seen that the Hollywood Commission for Eliminating Sexual Harassment and Advancing Equality was created in 2017, led by Anita Hill and backed by Lucasfilm President Kathleen Kennedy. This Commission is working with Hollywood companies such as Amazon, Netflix and ICM to develop a survey intended to get data on the abuse and power disparity that exists in Hollywood. Along with Times Up, which has raised over $22 million for a legal fund for women, there is the sense that this kind of social activism is helping to keep the goals of #MeToo in the public eye. Amy Baer, the president of Women in Film, noted that this kind of social activism has been successful and said, "What I do see is a consciousness that hadn't been there before and community building, particularly among women.... There's a lack of trepidation about speaking up and speaking out for people with powerful voices like actresses and prominent filmmakers, and that's been an amazing thing" (Faughnder and Perman 2020).

In terms of representation, finally, there has been a recognition of the need for more women to be directors and hold executive roles, and the fact that there were no women directors who were nominated for the 2020 Oscars, despite two very strong films directed by women, including *Little Women* (2019) and *The Farewell* (2019), demonstrates that more needs to be done on this front as well. While studios have slowly begun to include more women as directors, there are already some indications that women have begun to be given opportunities to direct big-budget franchises, including two Marvel Studio films, *Black Widow* (2020), directed by Cate Shortland, and *The Eternals* (2020), directed by Chloe Zhao. In stories as well, #MeToo themes are beginning to emerge, and films such as *Bombshell* (2019), based on the Fox News sexual harassment scandals, as well as *The Assistant* (2020), a fictional account that echoes the Weinstein case, suggest that the narratives being created on screen are also addressing the larger changes in the culture around finally acknowledging women's lived experience with sexual harassment. In the next chapter, we will see how these themes have emerged in storylines now coming out in films and television series. For the near future, however, the effects of the #MeToo movement on Hollywood continue to play out with mixed results for women who have come forward and for the industry that tolerated their mistreatment for too long.

2

Hollywood Storytelling in the Wake of the #MeToo Era

In an interview with the actress Sarah Snook, who plays the character Shiv Roy on the popular HBO television series *Succession* (2018–present), she was asked what it was like to be one of two women in the core cast of eight. She responded by saying that, as a result of the #MeToo movement, she was more confident about raising conversations with her co-workers about harassment (Rugendyke 2018). As she noted, "it's not a vilification of all men, it's about the behavior of certain men who don't treat women like humans … and it's the systematic behaviours that are allowed to be not acknowledged and allowed to go unchecked and unpunished" (Rugendyke 2018). As she reflected on the ways that harassment is manifested, she went on to define it, "On the greater scheme, it's sexual assault and everything else, and then on the smaller scale, it's just an off-handed comment about someone's appearance on the set, and how that may relate to your genitals or whatever" (Rugendyke 2018). She also described how she admired other Australian actresses such as Margot Robbie, who started her own film production company, and thought she would want to do something similar if the right project came along.

This description by Snook of how the #MeToo movement has made her more confident to speak up about these issues is instructive. Like many other actresses, her perspective signals that one of the ways that #MeToo has influenced Hollywood is through the conversations now taking place both behind the scenes as well as in the storylines of many recent shows, including *Succession*. A wide variety of television series and films have tackled themes around #MeToo. Emily Nussbaum has also observed that the #MeToo movement has caused a more general reflection on the ways in which morality has earlier been represented on television in terms of issues affecting women. Nussbaum points out that in episodes on television in earlier eras, such as when Edith Bunker fought off a rape attempt in a 1977 episode of CBS's *All in the Family* (1971–1979), it was treated as a "very special episode" (Nussbaum 2019). Sexual violence toward women during this time, when it was portrayed at all, was usually part of crime shows or soap operas. Sexual harassment in the workplace, on the other hand, was usually framed during these decades as humorous and as a way to heighten the romantic tensions between characters, as in CBS's *Cheers* (1982–1993), or the genial sexist comments by the boss Michael Scott on NBC's *The Office* (2005–2013).

The 2001 30th episode of HBO's *The Sopranos* (1999–2007), titled "Employee of the Month," marked a turning point in television in terms of the issue of sexual assault. In the episode, Tony Soprano's therapist, Dr. Jennifer Melfi, was violently raped by a stranger in the parking lot of her therapy office. The scene itself was very brutal with close-ups of Dr. Melfi's face as she is terrorized and screams out and cries. There was no music or sound effects or special camera work but a realistic focus on the terror of a woman being raped.[1] The episode was radical in many ways in portraying a woman's experience being sexually assaulted. For one, the series itself was focused on toxic masculinity, and several shows after *The Sopranos* continued the theme of male anti-heroes such as the character of Walter White on AMC's *Breaking Bad* (2008–2013) or Don Draper in AMC's *Mad Men* (2007–2015). While the female characters are an important part of the lives of the men on *The Sopranos*, including Tony's wife Carmela (Edie Falco), or his psychopathic mother, Livia (Nancy Marchand), the women are usually categorized, as television critic Anne Cohen has noted, as either wives, mistresses, mothers, daughters and "whores" (Cohen 2019). The character of Melfi, however, was not easy to categorize in the constellation of women. And, while the rape could be seen as another example of toxic masculinity, it was not used to further the male character's development or to pose Tony Soprano as someone who would rescue Melfi, because Melfi made the choice not to tell Tony why she looked beat up. In so doing, she knew that she was putting aside her desire for vengeance, which would have occurred if Soprano knew what happened to her because he would have tried to find out who attacked her and would have had him killed. Instead the episode showed a more realistic portrait of what happens when a woman is raped and the rapist is not caught. David Chase, the creator of *The Sopranos*, offered his description of the episode: "If you're raised on a steady diet of Hollywood movies and network television, you start to think, 'Obviously there's going to be some moral accounting here.' That's not the way the world works. It all comes down to why you're watching. If all you want is to see big Tony Soprano take that guy's head and bang it against the wall like a cantaloupe.... The point is—Melfi, despite pain and suffering, made her moral, ethical choice and we should applaud her for it. That's the story" (Reddit 2017).

More generally, this episode signaled a new era as plotlines on television routinely began to include sexual violence, and these assaults were oftentimes portrayed as also explaining the character's backgrounds and subsequent reactions to events. At the same time, television began to open up and have more women as creators and writers, which in turn allowed for more kinds of women's stories, which often included stories of sexual assault. One of the unfortunate by-products, however, of the increase in these stories is that there became a push to show ever more stories that were often highly sexualized. Because they were seemingly showing misogyny, little distinction was made for a viewing audience between seeing sexual violence and using it as a form of titillation. Recent shows like HBO's *True Detective* (2014–present), which portrays a female corpse who has been flayed, is one example of this kind of use of women's bodies for graphic portrayals, and as Nussbaum noted, "there wasn't much difference between what the camera ogled and what it critiqued" (Nussbaum 2019).

Other shows in the ensuing time period, such as HBO's *Game of Thrones* (2011–2019), also drew on these sexualized images of women being sexually assaulted and abused. These representations were similarly thought of as being artistically "serious" while at the same time capitalizing on the increased viewership gained through pornographic images of women being abused. There have been many similar renderings of sexual violence against women, particularly young, female victims in the wave of crime dramas on "quality television series," such as *The Bridge* (Sweden's *Sveriges Television* and Denmark's *Danmark's Radio,* 2011–2018), *The Fall* (Netflix, 2013–2016) and other "Nordic noir" shows imported from Northern Europe. The female corpse has become such a standard trope that it is invariably shown in opening episode of many different series and can be counted on as an easy form of shorthand for audiences who may not understand the larger culture of other countries but can easily understand the meaning of a dead, white, female corpse.

The #MeToo era, by contrast, is arguably bringing forward new stories about women and sexual harassment; while this could also be seen as beginning prior to the movement, it is reaching new heights in the wake of it. These include a comedy created by Phoebe Waller-Bridge, *Fleabag* (BBC Three, 2016–2019). Waller-Bridge plays a character named Fleabag, who is single and living in London. She is sexually active, and sexualizes almost every situation she finds herself in. Her most important relationship is with her woman friend Boo (Jenny Rainsford), though much of the season is a reflection on the loss of Boo and her reaction to it through acting out sexually and in self-destructive ways. As if to highlight the sense that this is a kind of precursor, or "before" time, to the themes that will be more explicit after #MeToo, at one point Fleabag says to her father, "I have a horrible feeling that I'm a greedy, perverted selfish, apathetic, cynical, depraved, morally bankrupt woman who can't even call herself a feminist" (Nussbaum 2016). Viewing this as one of the spate of "Bad-Girl" comedies in this year, 2016, Emily Nussbaum finds that, unlike many comedic attempts to make women as anti-heroic as men have been on recent television series, Waller-Bridge is much more of an original because she is both the author of the difficulties she gets into while at the same time a "figure of pathos" (Nussbaum 2016). Though *Fleabag* pre-dated the #MeToo movement her ability to portray the ways men act on their privilege and want to receive forgiveness for it, as well as the ways women are somehow morally "bankrupt" and can't even call themselves "feminist," signals how the show is struggling with the issues that will emerge more fully in the wake of #MeToo.

In the first episode ("1.1," first aired July 21, 2016), Fleabag is shown going for an interview and accidentally pulls off her sweater in front of a bank loan officer in the middle of an interview for a bank loan. She turns toward the camera and comments to the audience about the action in the scene. In another absurdist moment, she looks at the other people on the tube, or subway train, in London, and they look like they are laughing hysterically, but it also seems like Fleabag is hallucinating. She then turns to the camera and says to the viewers, "I think my period's coming." After Fleabag takes off her shirt by accident, a later episode (Season 1, episode 4, first aired August 11, 2016) shows the same bank loan officer attending a male retreat where he confronts his anger toward women. The episode begins when Fleabag and her sister Claire go to visit a female-only silent retreat that their father had paid for them to attend. While she is at the retreat, Fleabag sees the bank loan officer, who is at another workshop he was required to attend because of a sexual harassment scandal that had occurred at work. Both Fleabag and the bank loan officer bond over their status as attendees and because they are both unhappy.

The episode, which is played for laughs, is actually powerful in showing the rage that men feel toward women, as the men at the retreat are punching inflatable female dolls and screaming at them. It turns out that the bank loan officer was a sexual harasser who was sent to the

Fleabag (BBC3, 2016–19): Creator Phoebe Waller-Bridge, shown here with the bank manager (played by Hugh Dennis), broke new ground in her portrayal of a woman negotiating the aftermath of trauma.

weekend retreat to deal with his issues with women. By the end of the episode, however, he ends up in a kind of complicated resolution with Fleabag, and by the end of the series, he works with her to secure a bank loan for her failing café. Describing this show, Nussbaum noted that it was basically written *into* the #MeToo movement—but that like good art is able to do, it was channeling or "mirroring" the prevailing zeitgeist or anxieties of the era, "blinding fury at what men get away with and desperation for some path to forgiveness, along with an ugly awareness of how those two impulses might contradict each other" (Nussbaum 2019).

Speaking about the importance of the #MeToo movement in 2018, Waller-Bridge, who also wrote the female-centric BBC series *Killing Eve* (2018–present) which stars Sandra Oh and Jodie Comer, noted that the TV industry "can no longer get away with not having the conversation" about the topics raised by the "#MeToo movement" (Ravindran 2018). At the same time, Waller-Bridge is acutely aware of how difficult it is to use humor in a post–#MeToo world, because the world we are living in now is different in terms of how jokes might be heard. Waller-Bridge noted that, at first, there was relief that the men in the entertainment industry were being exposed for their behavior, but then at the same time, it made it difficult for her to write jokes about women's sexuality in the same way she did before. While there will always be "room for humor," all of a sudden a lot of the jokes "stopped being funny," but reflecting on the new conversations being held after #MeToo doesn't mean that you have to stop writing jokes about female sexuality (Newis-Smith 2019).

Broadcast television in the United States responded with stronger stories about women in the #MeToo era in much more force in the 2019 season. On channels as varied as NBC, CBS, ABC, Fox and the CW, there were several pilots, many of which were dramas and dramadies, that featured strong female characters, some of whom had #MeToo stories in their backgrounds. Though many of these shows did not make it to the fall line-up, it was clear there was a trend toward independent women who struggle against the odds. Interestingly enough, many of these shows were developed by CBS, whose board had fired Leslie Moonves, the CEO embroiled in a sexual-misconduct scandal. In one pilot, *The Republic of Sarah*, the main character, played by Sarah Drew (of *Grey's Anatomy*), was the mayor of a small town in New Hampshire that decides it wants to become an independent nation. In another pilot, a political thriller called *Surveillance*, Sophia Bush plays the head of communications at the NSA. In a third pilot, *Tommy*, Edie Falco (who played Toni Soprano's wife in *The Sopranos* and Nurse Jackie in *Nurse Jackie*), plays the first chief of Los Angeles' Police Department. On other channels in 2019, similarly, women were cast as assertive, leading characters in power. On NBC, there was a pilot called *Emergence* (2019–present), that starred Allison Tolman (of *Fargo*) who plays a sheriff who discovers a child after a mysterious accident. ABC, as well, had another woman in uniform, *Stumptown* (2019–present), starring Cobie Smulders (of *How I Met Your Mother*), who had been an Army vet and is now a private investigator. On the CW channel, there was a pilot for *Batwoman* (2019–present), starring Ruby Rose (from 2017 *Pitch Perfect 3*) who plays Batwoman, aka Kate Kane. The Fox channel also put forward a comedy called *Patty's Auto* (ND), starring Carra Patterson (from 2015 *Straight Outta Compton*) who runs an all-female garage of mechanics (Rice 2019).

In addition to the networks, streaming services as well as cable television created more #MeToo stories. On HBO, for example, several shows featured references to #MeToo, both directly and indirectly. These references were interwoven into storylines, in some cases as highlighting the backstory of a character and their inner conflicts, while at other times they were used to punctuate or offer a dramatic plot twist in a show that otherwise wasn't about sexual harassment or assault. The HBO series *High Maintenance* (2016–present) is a good example of the former. In this series, which focuses on the life of a character known as "The Guy," a marijuana dealer working out of Brooklyn, there is the Season 3, episode 7 "Dongle" (original air date March 3, 2019) where the Guy has a conversation with his ex-wife regarding their feelings about the ex-wife of an actor recently called out for sexual harassment. The Guy's ex-wife said she thought it was problematic that the abuser's own ex-wife defended him. The Guy then says that "[y]eah, but you and I both know that's a tricky situation," and his wife says, "Absolutely. No one's saying anything to the contrary." The Guy then says, "She left him," and the ex-wife responds, "after she made excuses for his behavior. That's all I'm saying." The Guy then responds, "The heart wants what it wants. Oh, that's a Woody Allen quote. Whoops" (Lemiski 2019).

This kind of ambiguity is used to illustrate how complicated the conversation around sexual harassment can be, and how the Guy is struggling with his own sense of whether the ex-wife was herself guilty in some way for leaving the harasser without condemning him. In the last episode of the third season, "Cruise" (original air date March 17, 2019), the Guy deals more directly with the issue of #MeToo as he talks to an old girlfriend from high school. He reminds her that in high school there was an evening where they were at a concert and she was wearing pajama bottoms and he pulled down her pants playfully, and she wasn't wearing any underwear. He said he has been carrying around guilt about that incident for the past 20 years, and then she tells him she doesn't even remember.

This storyline where the male character is dealing with his own internal working through of what #MeToo means is contrasted by other shows that use a #MeToo storyline to drive a plot. For example, in HBO's *Succession* (2018–present), which is loosely based on the Murdoch media empire and the titanic conflicts that take place between the patriarch and his children, the issue of sexual harassment plays a central role in the plot in the ninth episode of Season 2, titled "DC." The media empire, which is controlled by Logan Roy (Bryan Cox), has as one of its subsidiaries a cruise line where for some decades sexual harassment and worse was carried out by "Mo" Lester, who ran that division. Even though it is clear that there was knowledge of his harassment and other illegal acts, including possibly even a death, the family kept it under wraps. At the conclusion to the second season, though, the truth about this harassment comes to light and threatens to cause the Logan family the loss of their entire empire. This episode and others like it reflected in turn the deluge of plots about acts of sexual harassment that were written as a way to seem current with the events of the day. Other shows, in turn, such as *High Maintenance*, looked more at how people were somehow considering their own collusion with the system that accepted the kinds of acts that #MeToo finally called into question. Just as television is a mirror for the kinds of conversations going on in society, then, these shows, in various ways, reflected the ways society was confronting the fallout from #MeToo, whether it was through fictional characters examining their own and others' behaviors, or through looking at the ways those accused of sexual misconduct were being brought down in a variety of industries.

In addition to series on HBO, other streaming television shows also illustrated the impact of #MeToo in terms of their characters and plotlines. *The Boys* (Amazon Prime Video, 2019) was an absurdist superhero series where the creators described being influenced by the #MeToo movement. As showrunner Eric Kripke offered, "A couple weeks in, we all looked at each other and said 'Jesus Christ, I think we could be making one of the most current shows on TV' ... it became endlessly relevant the more we explored" (Clark 2019). The show, which premiered on Prime Video in July 2019, was co-created by Seth Rogen and Evan Goldberg and follows the workings of a group of mercenaries who try to bring in superheroes who use their power for bad ends. It is based on the comic book with the same name, written by Garth Ennis and artist Darick Robertson and put out by Dynamite Entertainment. The superhero characters are egotistical and owned by a corporation named Vought that in turn uses them to merchandise their products. The main character on the show is Hughie Campbell (Jack Quaid), who is recruited to work for Billy Butcher (Karl Urban) for his team called "The Boys" after there had been another tragedy involving a superhero (Clark 2019).

As the show was being conceived in 2015, Ennis, the writer of the comic book, told Kripke that he was interested in showing what would happen if you "combined the worst of celebrity with the worst of politics" (Clark 2019). While coming before Trump was elected, the series was seemingly prescient in its storyline about the toxic combination of media and politics. The show was then picked up just as the #MeToo movement was happening, which in turn caused the creators to re-think the sexual assault that was part of the original storyline. In the story, the superheroes welcome a new member of the team, a woman named Starlight (Erin Moriarty), by sexually assaulting her as a group. Homelander (Antony Starr) is the superman hero who drops his pants and demands a blowjob from Starlight, and then two other superheroes, Black Noir (Nathan Mitchell) and A-Train (Jessie T. Usher) also arrive and demand the same from her. The character Annie/Starlight gives in to their demands, but she is traumatized and ashamed (Outlaw 2019).[2]

By the time they began filming the series, Kripke was at first contemplating leaving out the Starlight sexual assault subplot, but the writing staff challenged him on this. As he explained, "This was my female writer and producer saying, 'This is something that happens, we think it's important to talk about'" (Outlaw 2019). In the first iteration of the show, before the Weinstein scandal, there was a tamer version of the incident where the Starlight subplot looked more like a female who was trying to work in an all-male environment where she had little protection against being sexually harassed. However, when the Weinstein scandal broke in late 2017, Kripke felt that the massive shift in the cultural conversation around sexual assault made him re-think the central scene again. Describing this process of revising it, Kripke offered, "I was horrified. But the books play it out as shocking and I said, if we're going to adapt it, let's present it with the greatest responsibility and present it as the horror movie that it is. There was a lot of debate, balancing, and angst right up until shooting. When Erin Moriarty [who plays Starlight/Annie January] came on board the show, we had a long conversation with her about it. We polled everybody because we wanted to make sure we were getting it right. I was nervous as s---! It's the most serious thing I've ever done. So I wanted to get it right" (Clark 2019).

As the storyline unfolds in the rest of Season 1, there are continuing echoes of the #MeToo movement, including how the public reacts to Starlight opening up about her assault as well as a storyline about how the evil Vought company tries to subvert her efforts through their own PR stories. In these ways, television series like *The Boys* made a conscious decision to alter or change their original storylines once the #Me Too movement broke, both as a response to the events that were occurring in real time, as well as because they wanted to maintain their relevance in the face of a new reality where women's stories of sexual assault were finally being heard and believed.

Reactionary Stories in the Era of #MeToo

The stories being created in the era of #MeToo in film and television are progressive in some cases, while in others they are arguably more reactionary. In some instances, the reactionary plotlines are created specifically to make a point about the need to take the #MeToo movement seriously and use role reversals to highlight this. This is the case in Episode 2 of the final season of HBO's *Veep* (2012–2019) "Discovery Weekend" (original air date April 7, 2019). This storyline shows women in Washington, D.C., trying to distance themselves from a particularly obnoxious character named Jonah Ryan (Timothy Simons), who is running for office, and they create a hashtag called #NotMe—to signify a movement where women stand up to say they never dated Jonah Ryan. The women were tired of him bragging about the women he had supposedly slept with, and one woman (played by Heidi Gardner of *Saturday Night Live*) asks him to sign a non-disclosure agreement to get him to stop claiming they had been romantically involved. She says that they only went on one professional lunch and that they had split the bill. He tells her, "There's no way I'm gonna sign that," and "You are a stone-cold neck-down hottie, and I want the whole world to know we dated" (Maple 2019).

After he refuses to sign the agreement, she goes to the media with her lawyer in tow, and the lawyer tells the media that women have "been silent in the face of rumors they went out with Congressman Ryan" for too long. The young woman then makes the following statement to the media: "Jonah Ryan and I have not ever dated nor gone on a date of any kind. We had one meal together, but it was strictly professional, and in the presence of others. It was really more of a group thing. He once tried to friend me on Facebook, and I did that thing where I never responded 'yes' or 'no.' Hoping—praying—that it would end. His behavior was completely appropriate at all times" (Maple 2019). Describing how they came up with the storyline of #NotMe around Jonah and of the seventh season in general, the showrunner David Mandel offered: "Obviously, things were going on in the world and you're not writing the show in a box. We didn't sit down and say, 'Today's the day we're coming up with a #MeToo storyline.' But it was certainly something we wanted to address. Even back to the Selina thing, there's a #MeToo aspect to her not being bothered by having her ass grabbed in the past. It's in the air and it definitely exists in the season in general, because this is what's going on" (Strause 2019).

By contrast, some storylines were arguably reactions against the challenges posed by the #MeToo movement. For example, in *The Romanoffs* (Amazon Prime Video, 2018) created by *Mad Men*'s (AMC, 2007–2015) Matthew Weiner, there was a storyline that explicitly touched on issues of sexual harassment and abuse. The release of the series was itself arguably part of a larger cultural moment where questions were raised about whether men who had themselves been accused of sexual harassment should be allowed to return to their jobs in the entertainment industry. Some of these men included the comedians Louis C.K. and Aziz Ansari, who went on to perform stand-up comedy sets after they had been accused of misconduct, while other actors such as James Franco and Jeffrey Tambor went on to appear in new seasons of their shows, *The Deuce* (HBO, 2017–2019) and *Arrested Development* (Fox,

2003–2006; Netflix, 2018–2019), respectively. In 2017, Weiner was accused of sexually harassing Kater Gordon, a writer on *Mad Men*. Gordon said there was an incident where he demanded to see her naked. Weiner denied that he engaged in any inappropriate behavior with Gordon. Yet, in the third episode of *The Romanoffs* titled "The House of Special Purpose," he chose to portray abusive behavior as part of the storyline, though he claimed that he had in fact written the episode before he was accused of anything.[3]

The story begins with an actress named Olivia (Christina Hendricks, who also starred in *Mad Men*) who goes to Austria to make a miniseries about the Romanoffs, a family of Russian Royals. While filming their story, the actress is abused by the cast and crew, both sexually and physically, and at the end of the show, a final act leaves her dead. The primary person abusing her is Jacqueline (Isabelle Huppert), an actress who became a director, portrayed as a crazy older woman who torments and mocks Olivia and encourages other people on the set to do the same. In one of the scenes, another male co-star attacks Olivia and says that the sexual assault, while not in the script, was "in the moment." When Olivia calls her agent to plea to leave the scene, he makes fun of her and tells her that she might be perceived as a "difficult woman" if she leaves. In the last act, Olivia is kidnapped and put into a room with the other actors, where she is led to believe that real bullets are flying around her as the other actors around her fall down and are bleeding. Though it is supposedly a ruse in order to get her to perform more realistically for the big Romanoff death scene that occurs, it turns out that she is so scared by it that she actually dies of fright.

In describing her understanding of the role and the way in which the actress is portrayed, Christina Hendricks defended the narrative of sexual harassment that was part of the storyline as a literary vehicle to highlight the problem of sexual harassment. As she explained, "It's natural that [sexual harassment] would come into *The Romanoffs*. A female actress is going to experience this kind of treatment and this kind of behavior all the time. And it seemed appropriate too, when this woman is dealing with sort of a mind-game situation as it is, that all of these elements would play into her insecurities" (Hallemann 2018). For television critics such as Kelly Lawler, however, the episode is exploitive, and Olivia's anguish is used for "shock value and cheap laughs" (Lawler 2018). Noting earlier Hollywood films, Lawler makes the point that this kind of portrayal of women's trauma is routinely used, or as she pointed out, "even before #MeToo, Hollywood's track record in portraying women onscreen is spotty at best. Dating back decades (from *King Kong* to Alfred Hitchcock and Woody Allen films), female characters have been put through constant maltreatment in stark contrast with their male counterparts on the big and little screens" (Lawler 2018). And, though there was time between Gordon's accusations of sexual harassment against Weiner and the airing of "The House of Special Purpose," neither Weiner nor any of the executives at Amazon responsible for the series chose to delay or stop the episode from airing.

Whether Weiner was intentionally showing a woman being traumatized for shock value or not, the echoes of #MeToo in reference to his own life are more clearly delineated in the fifth episode of the first season, "Bright and High Circle" (original air date, November 2, 2018). The title comes from a line in a poem by Alexander Pushkin and is read early on in the episode in the scene of a college course on Russian literature. The poem begins with "When you're so young and fairy years/Are smeared by the gossip's noise,/And by the high world's trial, fierce,/Your public honor's fully lost;/Alone midst indifferent crowds,/I share with you your soul's pains" (Gilbert 2018b). The poem, about someone who is slandered, resonates not only with the theme of the episode, but arguably with Weiner himself, who was accused of sexual misconduct.

The episode, which was co-written by Weiner and Kriss Turner Towner, revolves around a piano teacher, David Patton (Andrew Rannells), who works for several wealthy families in Los Angeles teaching their children how to play piano. Unlike the earlier story where a woman is mistreated, in this story it is a man who is subject to innuendo and gossip when an accusation is leveled against him that he acted inappropriately with one of the students he was teaching. When the piano teacher's primary patron Katherine (Diane Lane) learns that he was accused of sexual misconduct, she panics because her three sons have been having lessons with him for years. She contacts her other friends whose children also have had lessons with him, which in turn sets off more suspicion and fear about whether he acted inappropriately with their children. Even though her own children love David and are adamant that he didn't do anything improper with them other than tell an off-color joke, Katherine is still suspicious, which only intensifies when it is revealed that David lied about being a descendent of the Romanoffs, which she actually is.

Katherine and her children are admonished by her husband that the worst thing you can do is ruin another person's reputation with a false accusation. The father forces them to continue taking piano lessons with David, saying, "When you accuse somebody of something, whether they did it or not, you make everybody look at them differently.... Bearing false witness is the worst crime you can commit. Otherwise, anyone can say anything about anybody, and just saying it ruins their life. No

The Romanoffs (Amazon, 2019): Writer-director Mathew Weiner responded to an accusation of sexual harassment in the episode "The Bright and High Circle." Shown here are Katherine Ford (played by Diane Lane) and David Patton (Andrew Rannells).

matter what they did. Does that seem fair? It's not fair" (Gilbert 2018b). At the end of the story, David is allowed to keep teaching and nothing happens to him, except his reputation has come into question.

If "Bright and High Circle," then, feels like an extended response to Weiner's own sense that he had been unfairly accused, it also echoes references from other men who started to come forward during this time and responded that they too had been wrongfully accused of sexual misconduct. These men, including John Hockenberry from *National Public Radio* who wrote an essay for *Harper's* magazine, as well as Jian Ghomeshi, who wrote a defense of his actions in the *New York Review of Books*, together represent a response where men who are accused of sexual harassment write about the impact of being accused and its effect on them. As we will see, for many men who had been accused of sexual misconduct during the #MeToo era, there is the sense that these accusations felt like a kind of witch hunt. In addition, there tends to be a focus on the feelings of the men themselves in these responses, rather than the women who accused them of misconduct.

To summarize, in the aftermath of the first accusations against Harvey Weinstein to the time when he was sentenced to 23 years in prison for rape and sexual assault (March 2020), the cultural movement which led to his downfall could be seen in a wide variety of film and television series. The themes of workplace ethics and sexual misconduct in the workplace, and the so-called "casting couch" culture have been portrayed in fictional and nonfictional stories that Hollywood has turned out with increasing frequency. In some stories, there was an emphasis on the character's reflection on their own behavior, and in others, the focus was on the reactions by the individuals who were harassed. Another representation was of men who were not necessarily monsters or predators, and women who were not simply victims. Rather, there was an effort to offer more nuanced portraits of the dynamics of sexual harassment and assault. Another representation, finally, could be considered reactionary in the sense that the males themselves were portrayed as somehow victimized by a culture that had gone too far in its accusations of sexual harassment.

Thinking about the array and cultural importance of these representations, media scholar Robert Thompson has noted, "Hollywood is now becoming its own loudest voice in helping to call out what a bad thing this is" (Serjeant 2020). In his view, while the #MeToo movement was instrumental in helping bring these issues to light, it becomes "institutionalized" when they are put into film and television series that people will continue to watch for years after the news coverage of #MeToo fades (Serjeant 2020). At the same time, some of these stories were arguably reactionary and spoke to a sense of injustice on the part of creatives who had themselves been challenged about their own behavior. In the next chapter, we will look more closely at how the stories around #MeToo are played out in reactions by the harassers themselves, and also how they reveal the complicated set of responses from the industry to the issues raised in the wake of the movement. Sometimes angry and reactionary, these responses highlight the challenges faced by an industry whose own practices and actions were being called into question at the same time.

3

The Backlash Against #MeToo and the Quest for Redemption

While there have been changes in the kinds of stories being told in both U.S. and international film and television series, there have also been changes in what is considered appropriate and inappropriate behavior in the industry itself. At the same time, a new development has occurred which calls into question the very premise of the #MeToo movement. These counter-narratives take different forms, ranging from the idea that there are no distinctions being made between different kinds of acts that are considered harassment, to the charge that the #MeToo movement is a "witch hunt" bringing down men who have not been charged and convicted with anything in a court of law. One example of this kind of thinking is from none other than Roman Polanski, who was himself accused of sexual assault against a minor and who criticized the movement as a form of "collective hysteria of the kind that sometimes happens in the society" (Sharf 2018). He told a Polish edition of *Newsweek* magazine that he believed people are endorsing the movement out of "fear," and that they are afraid if they don't, they will be ostracized. Polanski ended his interview by comparing the supporters of the #MeToo movement to people who mourn their leaders in North Korea who cry when they die, that "you can't help laughing" (Sharf 2018).

One of the primary criticisms of the #MeToo movement Polanski has touched on is the idea that it has somehow gone too far. This criticism has been repeated in any number of narratives and has been voiced in both popular media as well as in television and print interviews with entertainment figures. One example is an interview Ronan Farrow did on *The Bill Maher Show* on HBO (April 27, 2018). When asked by Maher whether the #MeToo movement has gone too far, citing former comedian and Senator Al Franken and Aziz Ansari as examples of men who were brought down and had to resign (in the case of Franken) or lose their television show (in the case of Ansari), Farrow replied that the culture was pretty good with "self-regulating." As for Ansari, it was a single source narrative about a date gone wrong, but no one saw him as Harvey Weinstein. Maher responded by saying Ansari did suffer a lot from the publication by the young woman of a "date gone wrong," because he lost his television show contract. Farrow responded by noting that the vast majority of reporting on sexual harassment and assault has been meticulous and has responded to serious crimes. Maher then asked about Al Franken and how he felt compelled to resign from the U.S. Senate, and Farrow responded by noting that it is important to make these kinds of distinctions between different acts moving forward. At the same time, these conversations have been silenced for decades. It is important that it is coming out now, even if it seems like it is coming out in torrents (Karlis 2018).

The response by Maher is indicative of what would become a growing reaction to what has been perceived as the excesses of the #MeToo movement. Bari Weiss (2018), for example, wrote an article titled, "Aziz Ansari Is Guilty. Of Not Being a Mind Reader" in which she claimed that being a sexually active woman in the 21st century means, according to those who would accuse Ansari of wrong doing, being a victim of sexual assault. In her view, the #MeToo movement has transformed what should have been a movement for women's empowerment into a symbol of "female helplessness." She then recounts the young woman's accusation that after Ansari undressed her, he persistently kept trying to have penetrative sex with her and that when she tried to voice her hesitation, he ignored her. When she finally uttered "no" he shifted and said, "How about we just chill, but this time with our clothes on?" (Weiss 2018).

According to the young woman, he ignored her "verbal and non-verbal cues," but for Weiss, who asserts that she herself is a "proud feminist," the fact that the woman was "hanging out naked with a man (means) it's safe to assume he is going to try to have sex with you" (Weiss 2018). In her view, while Ansari may have been aggressive or selfish, it was up to the woman to be more verbal and that as the woman tells the story, Ansari is the only one who has any "agency" in the encounter. For Weiss, the current form of feminism that the #MeToo movement represents treats the woman as if she has no agency, and so she was the subject of a "flagrant abuse of power" in the sexual encounter. However, to equate this encounter with what happens to actresses who are sexually assaulted or working class women whose bosses demand they have sex with them is to "trivialize" the #MeToo movement. And while it is important to try to socialize young men so that they are not equating what they see in pornographic videos with how they should behave, as well as to teach young women to speak up more about their needs and desires, Weiss believes it is wrong to criminalize "awkward, gross and entitled" sex by men (Weiss 2018).

Daphne Merkin (2018) also challenges the ways in which the #MeToo movement has somehow moved from a "bona fide moment of accountability into a series of ad hoc and sometimes unproven accusations." In her view, too many people are being accused in an overreach of "political correctness" which has become itself a form of "social intimidation" (Merkin 2018). In this climate, Merkin believes that many women, including feminists, are secretly thinking that these acts don't constitute sexual harassment but are afraid to speak up and that the kind of sexual harassment men like Franken or Ansari are being called on is just a part of real life.[1] For Merkin, while Matt Lauer and Kevin Spacey clearly engaged in criminal acts, there are countless other accusations that are more scattered, or "vague and unspecific," against people like Al Franken, Ryan Lizza, Jonathan Schwartz or Garrison Keillor.

Echoing the arguments of Bari Weiss, Merkin believes that one of the most troubling aspects of this new climate is that younger women are now being viewed as they were during the Victorian era—as frail and in need of protection. She cites the example of a campaign to remove a painting by Balthus from the Metropolitan Museum of Art because it showed a young girl in a suggestive pose. While this campaign was ultimately unsuccessful, Merkin compares this attempt at censorship to similar acts of censorship by religious zealots in earlier eras (Merkin 2018). One of the arguments, then, that is being launched against the #MeToo movement is that it has taken away women's sense of agency that would allow them to say "no" to someone who is harassing them. While she acknowledges that saying no might be risky, especially if you are saying no to someone who is in a position of power relative to you, she and other critics believe that the current atmosphere of denouncing people is not the way to solve it, and instead has created a climate which denies due process to those who are accused. A second problem for Merkin is that it's not entirely clear what the men are being accused of. She makes the distinction herself between being sexually harassed with "a degree of hostility," versus someone who is trying to kiss someone "in affection, however inappropriately," and ties this into a deeper confusion or ambivalence around expectations of behavior. At worst, she says that this "re-moralization of sex" has led to a kind of "inquisitorial whiff in the air" where people's fantasies will lead to them being "torched" (Merkin 2018).

In 2018, a year after the #MeToo began, there was an increasing number of articles that made arguments similar to that of Merkin and Weiss—that the #MeToo movement had gone too far. Catherine Deneuve, the French film actress, for example, had signed an opinion piece along with 99 other women in the French newspaper *Le Monde* that essentially argued that men should have the freedom to flirt. While they acknowledged that, yes, rape is a crime, "hitting on someone insistently or awkwardly is not an offense, nor is gallantry a chauvinist aggression." Deneuve and the other signers were responding to the French version of the #MeToo movement, called #BalanceTonPorc ("squeal on your pig"). This movement has called for laws around street harassment as well as a call to extend the statute of limitations for assault cases that happened when individuals were minors. Like Deneuve, the women who signed the letter were primarily professionals in the entertainment industry, many of whom were from an earlier generation and who wanted to challenge the calls for changes in the laws around sexual harassment.

Actors in the entertainment industry from the U.S. also began to speak out against the #MeToo movement. Sean Penn, for example, offered in an interview that he thought the #MeToo movement had been "largely shouldered by a kind of receptacle of the salacious" (Sharf 2018b). When he was asked to clarify what he meant by that, he continued, "Salacious is as soon as you call something a movement that is really a series of individual accusers, victim, accusations, some of which are unfounded.... The spirit of much of what has been the #MeToo movement is to divide men and women" (Sharf 2018b). Penn made a similar charge to Merkin and Weiss that the movement has conflated a wide variety of behaviors under one umbrella term, "sexual harassment," and that many of the charges are simply "unfounded." In his view, there is a lack of "nuance" in the movement, and as such, it has become "too black and white." In short, he believes that the movement needs to "just slow down" (Sharf 2018b).[2]

The idea that the #MeToo movement has somehow divided men and women and is too "black and white" became a common theme among people who somehow felt aggrieved at the speed and number of people who were ultimately accused of harassment. These critics believed that the movement was painting too many behaviors with the same brush and had gone too far in labeling them all as forms of sexual harassment. Liam Neeson, for example, used the term "witch hunt," to describe what he believed were the excesses of the movement and that too many men were being called out for what he believed were relatively minor offenses (Nordine 2018a).[3]

Like Neeson, other people in the entertainment industry similarly weighed in on the idea that the #MeToo movement had gone so far as to be a kind of "hysteria" that has resulted in ruined reputations and careers. Michael Haneke, the two-time Palme d'Or winner, and one of the most respected filmmakers in the European film community (he won the top prize at Cannes for two films: *The White Ribbon* [2009] and *Amour* [2012]), also criticized the #MeToo movement, finding that the move to judge people ended up with men having their careers ruined, which he described as "absolutely disgusting" (Nordine 2018b). Likening it to being convicted before being proved guilty, Haneke felt that the #MeToo movement was based on a "blind rage" that ended up destroying the men who were accused of harassment and concluded that the rush to judgment "destroys the lives of people, whose crime has not been proven in many cases ... this new man-hating puritanism that comes in the wake of the #MeToo movement worries me" (Nordine 2018b). And, as if to emphasize his sense of men being the victims of a witch hunt, he offers that men should not even "touch the topic" at this point in time. Like others who have charged that the #MeToo movement has gone too far, Haneke is quick to distinguish between sexual and violent assaults, which he believes should be condemned, from those who claim they have been sexually harassed, which he has likened more to a "witch hunt [that] should be left in the Middle Ages" (Nordine 2018).

These attempts to say that the movement has "gone too far" have been viewed by many as a way to either discredit the movement or, at best, as a misguided attempt to understand something by people who simply don't get it. Laura Dern, an actress who has worked for decades in Hollywood, believes this kind of criticism is based on a "lack of compassion" and that the women who are coming forward aren't simply accusing others of harassment for saying their "dress is pretty" (Erbland 2018). In addition, Tarana Burke, who initiated the use of the hashtag #MeToo, also pointed out that the backlash itself is a distraction, and that the purpose of the movement was to focus on allowing women and men who had been harassed and assaulted to tell their stories. As she noted, "Here's the reality: nine months ago, millions of people across the world raised their hand and said, 'This happened to me, too.' That's what they're saying ... they didn't call for anything. Even the women who came forward about Harvey Weinstein, they didn't call for him to be taken down, they didn't even think it was possible! But, at their own risk, they came forward to tell their truth" (Erbland 2018).

Reverse Victimhood: It's a Witch Hunt!

Some of the loudest voices against the movement have come from the men who have been accused. These men claim they were the real victims of the #MeToo movement, as opposed to the women who had come forward to tell their stories. In this role reversal, several men attempted to turn the accusations against them on their head and have claimed for themselves the role of victim at the hands of women who were falsely accusing them of sexual acts that they believed had been consensual. For example, in 2017, Matt Lauer, one of the co-anchors of NBC's *The Today Show* (1952–present), was fired after an NBC employee accused him of sexual misconduct. From her initial account, additional accusations of sexual misconduct from other women came out which contrasted sharply with the kind of friendly demeanor and banter Lauer had made his signature in his many years hosting this morning talk show.

In October 2019, Ronan Farrow, in his book called *Catch and Kill* (2019), made a very damaging allegation against Lauer. In the book, Farrow reported that the woman who had accused Lauer earlier, and who had been anonymous at the time in 2017, went on the record and accused Lauer of anally raping her without her consent. The encounter was excruciatingly painful, and Farrow quoted her as saying, "'It hurt so bad. I remember thinking, is this normal?' and that she eventually stopped saying no when Lauer continued to assault her, but wept silently into a pillow" (Pompeo 2019). Farrow's book was a huge indictment of the management of NBC News, which didn't respond to the allegations. Farrow himself had been working on the Harvey Weinstein story for NBC, but because he was continually thwarted in his investigation by his superiors there, he eventually went on to publish his story with *The New Yorker* (October 10, 2017, issue) magazine. When Farrow left NBC over their efforts to bury his Weinstein story, he then went on to make the connection between NBC's attempt to suppress this story about sexual harassment and the more general way that NBC News tried to downplay Lauer's sexual misconduct.

In the face of these new allegations, Lauer himself decided to launch a counterattack. He wrote a 1,400-word open letter that was released to news outlets by his legal team, in which he attempts to portray the young woman who accused him, Brooke Nevils, as a jilted ex-lover who was trying to punish him after he rejected her continued advances. In his view, he said that he was the real victim in the wake of the #MeToo movement. He attempted to undermine her credibility by saying, "She says I was the one pursuing the relationship, yet once it was over, she was the one calling me asking to rekindle it. She said she just wanted NBC to 'do the right thing,' yet she sought a monetary payment, and two years after I was fired, she is stepping forward to do more damage" (Pompeo 2019). In Farrow's account, Nevils acknowledges that the sex had been transactional, and that it was not a "relationship," but that she was "terrified about the control Lauer had over her career" (Pompeo 2019). Lauer, in trying to defend himself, was attempting to influence public opinion in his favor by asserting that he was simply doing the right thing, as a victim himself, to speak up about his treatment by the woman. Other critics have read this kind of self-defense of posing as the victim as a kind of backlash against the aims of the #MeToo movement in helping women come forward with their stories. Rebecca Traister, for example, has noted that the letter that Lauer published "concludes with the most backward powerful-man-as-real-victim bullshit I have yet read, and I have read a ton of that bullshit. He's presenting himself as the one who was silenced, who has suffered, who has had to do the hard work of speaking up—to loved ones—while his accusers have somehow walked free of responsibility or repercussion" (Pompeo 2019).

In the letter, Lauer doesn't assume any culpability on his part and Farrow discovered that many people at NBC also did nothing about it, despite the fact that they knew at the time what the allegations were. Disputing having had any prior knowledge about the assault, Andy Lack, the chairman of NBC News, firmly denied that NBC knew anything before the woman came forward. Countering this, Nevils responded to Lauer's letter, pointing out that it was a "case study in victim blaming.... I am not afraid of him now, regardless of his threats, bullying, and the shaming and predatory tactics I knew he would, and now has, tried to use against me" (Pompeo 2019).

Contrition

Other versions of the harasser-as-victim scenario are expressed in more benign terms by several celebrities than the arguably scorched earth tactics of someone like Matt Lauer. Casey Affleck, a film actor who also won an Oscar for his work in the film, *Manchester by the Sea* (2016), had previously settled two lawsuits in 2010 which had accused him of committing sexual harassment while on the set of his mockumentary film called *I'm Still Here* (2010). Though Affleck described himself as a supporter of the #MeToo movement, and believed that anyone would be supportive of it, he also felt that it was difficult for him to talk about the movement and that it "scared" him to discuss topics related to #MeToo, and that his mother raised him to be non-sexist (Sharf 2019b).[4]

Because of this, Affleck continued, "The way that I'm thought of sometimes by certain people recently has been so antithetical to who I really am that it's been frustrating ... and not being able to talk about it has been hard because I really wanted to support all of that, but I felt like the best thing to do was to just be quiet so I didn't seem to be in opposition of something that I really wanted to champion" (Sharf 2019b). Describing himself as misunderstood by others, Affleck tried to explain the context for the mockumentary where he was accused of sexual harassment, and he admitted that there was some unprofessionalism on the set of the film. There was a lot of "partying" and, because it was supposed to be a documentary, the harassment that occurred was also intentional as it was part of the movie. He believed that some of the crew members may have been confused by this and therefore didn't know how much "they were part of the movie.... It was a big mess and it's not something I would do again" (Sharf 2019b).

Believing he learned his lesson from this, Affleck vowed that he would be much more conscious and sensible in the future when he was on movie sets. In this scenario, Affleck expressed both anxiety that he would be misunderstood talking about #MeToo issues as well as feeling like he was only trying to make a film where his intentions were misunderstood. He didn't intentionally harass anyone, in other words, and in this way, his denial takes the form of having been misunderstood, something for which he does take responsibility. Interestingly enough, in his most recent effort as a director for a dystopian drama called *Light of My Life* (2019), the plot centers on Affleck, who stars as a father trying to protect his daughter after some unknown plague has killed the rest of the women in the world. There are some who believe that this kind of film plot is a reaction to the harassment lawsuits he settled, which he has denied at the film's premiere at the Berlin Film Festival in August 2019 (Sharf 2019b). Defending himself from these responses, Affleck reveals another aspect of the harasser as victim, that is, believing that critics won't approach his future work with an open mind and "be responsible and measured in their reactions" (Sharf 2019b).

Making Excuses for the Accused

Other actors and creative talent have had to grapple as well with the misconduct of the people they have worked with, and are similarly ambivalent about the behavior of these aggressors. In some cases, however, there is a sense that they are not sure whether these bad actors, so to speak, should be removed from their shows or forgiven and allowed to remain. In one show, *Arrested Development* (Fox, 2003–present;

Netflix, 2018–2019), Jeffrey Tambor, one of the lead actors, was accused of sexual misconduct on another show, *Transparent* (Amazon Prime Video, 2014–2019), but continued to work on *Arrested Development.* *Arrested Development*, which was created by Mitchell Hurwitz, is an absurdist take on a narcissistic family named Bluth who are in a family building business together. When sexual accusations against Tambor emerged from his behavior on *Transparent*, for which he was then fired, the ensemble cast of *Arrested Development* expressed their support for Tambor nonetheless (Deb 2018a). In a collective interview with the cast of *Arrested Development*, a wide-ranging discussion took place over the behavior of Tambor on the set of *Transparent* (Deb 2018a). One of the things he had to be forgiven for, according to Jessica Walter, who plays Tambor's wife on *Arrested Development*, was the way he treated her on set, at one point screaming at her. Walter, a 77-year-old actress, while in tears, offered "In almost 60 years of working, I've never had anybody yell at me like that on a set and it's hard to deal with, but I'm over it now" (Deb 2018a). Jason Bateman, who plays their son, then described Tambor's behavior as not out of the ordinary for certain performers, while Alia Shawkat, who plays Bateman's niece, then offered, "But that doesn't mean it's acceptable" (Deb 2018a).

As though to try to further explain what he meant, Bateman said it was normal to have these kinds of interactions where some cast member screams at another one, and said "But this is a family and families, you know, have love, laughter, argument—again, not to belittle it, but a lot of stuff happens in 15 years. I know nothing about *Transparent* but I do know a lot about *Arrested Development*. And I can say that no matter what anybody in this room has ever done—and we've all done a lot, with each other, for each other, against each other—I wouldn't trade it for the world and I have zero complaints" (Deb 2018a). Another actor from the show, David Cross, offered that Tambor should be applauded for saying that he learned from the experience, and that most people in his position have not said they have learned anything from it.

When the reporter asked whether a person who has admitted that they routinely yell at co-workers should be re-hired, Tambor himself said yes, if they say they have "reckoned with this" (Deb 2018a). In other words, Tambor defends himself by offering that, while he admits he has a temper and that he yells at people, he is working on it and because he has apologized for it, he should be re-hired. Bateman again came to his defense, this time explaining that the entertainment industry is characterized by people who are "in quotes, 'difficult,'" and that you need to contextualize those characterizations, whereupon Shawkat again challenged Bateman and reminded him that it doesn't mean it's acceptable

Arrested Development **(20th Century–Fox, 2003–06): Jessica Walter said her costar Jeffrey Tambor routinely demeaned her on the set.**

to behave that way. Tony Hale, another actor on the show, however agreed with Bateman, noting that "we've all had moments" (Deb 2018a). Walter then tried to defend her own reaction, saying that she needs to let go "of being angry with him," but that when Bateman jumped in and began to excuse Tambor by saying it happens all the time in the entertainment industry, Walter countered that in her 60 years of working in the industry she was never treated like that.

It's helpful to discuss this group interview in depth because it reveals the complex responses people in the entertainment industry have to those who have been accused of bad behavior on set, including yelling and screaming as well as sexual harassment. Interestingly enough, it was only when there was a strong and negative reaction on social media to Bateman's comments did he re-think his responses in the interview and apologize to Walter (Deb 2018b). When confronted with the angry reactions to his interview defending Tambor, Bateman responded by saying he was "embarrassed" and "sorry" that he did that to Walter and that it was a "big learning moment" for him as well.[5] For many viewers, however, this learning curve by Bateman doesn't really address what happened, as one reader noted that Bateman felt compelled to defend Tambor not once, but nine times during the interview even after Walter is reduced to tears thinking about how much Tambor screamed at her (Deb 2018b).[6] The larger point is that there is still an impulse by many people in the entertainment industry to normalize or make excuses, or "contextualize," what is clearly abusive behavior both because it is easier to excuse it if it hasn't happened to them as well as because it is profitable for the industry to allow these actors to remain on the shows.[7]

Other actors and directors and men who work in the entertainment industry have found that another tactic to diffuse criticism of their past actions, or the actions of those they have worked with, is to pre-emptively apologize for *past* behaviors that might be considered abusive in the *current* climate. Directors such as Quentin Tarantino, for example, have pre-emptively apologized for past behaviors as a way to circumvent any potential criticism. In this way, Tarantino and others have been able to remain in the industry despite accusations lodged against them about being complicit with or knowledgeable about the behaviors of serial harassers as well as acting aggressive themselves in their professional lives. Olivia Munn, an actress, specifically called out Tarantino, noting that "[w]e have Tarantino who admitted to abusive behavior on set and also knowing what Harvey Weinstein was doing" (Sharf 2019a).

The case of Tarantino is especially instructive, both because Tarantino himself was closely tied to Harvey Weinstein as well as because he represents an attempt made by many individuals in Hollywood and beyond to come to terms with their past complicity and behavior in a way that leaves little room to further criticize them. By admitting some culpability, in other words, either as a bystander to other harassers and not doing anything about it or admitting their own bad behavior, these individuals hope to pre-emptively downplay any further attempts to hold themselves or those they worked with accountable for their actions. Tarantino, for one, has said that he feels ashamed that he didn't speak up sooner about Weinstein and that he continued to work with him after he heard about his behavior toward women. As he explained to Jodi Kantor, one of the reporters who broke the story about Weinstein for the *New York Times*, "I knew enough to do more than I did … there was more to it than just the normal rumors, the normal gossip. It wasn't secondhand. I knew he did a couple of these things" (Kantor 2017).

Because of his knowledge, Tarantino said he should have taken responsibility for what he had heard at the time, and if he had, he would have not been able to work with him. The specific cases that he knew about and which he didn't respond to when he heard about them had come directly from the women involved, including Mira Sorvino, who was in a relationship with Tarantino at the time Weinstein's harassment of her occurred. He also heard a similar story of unwanted advances several years later from the actress Rose McGowan, who had reached a financial settlement with Weinstein. Part of the reason that Tarantino claimed he didn't act sooner was that he said that he felt that the stories didn't really constitute a larger pattern of abuse by Weinstein. In the meantime, he continued to make films with Weinstein, who in turn continued to champion Tarantino and was instrumental in Tarantino's own professional opportunities and success. Tarantino admits to having minimized Weinstein's behavior as inappropriate but mild and admitted that "[a]nything I say now will sound like a crappy excuse" (Kantor 2017).

In fact, Tarantino and Weinstein had a very close working relationship for decades, with Weinstein producing many of Tarantino's films from 1992 on. Weinstein also threw Mr. Tarantino an engagement party in 2017. When he read about accusations against Weinstein, including rape, Tarantino described himself as being "horrified" (Kantor 2017). In 1995, however, when he was dating Sorvino, he did admit that he heard about his harassment of her, but thought at the time it was just because Weinstein was particularly "hung up on Mira" (Kantor 2017). He also heard other accounts through the years, but minimized them, and has said that he now regrets not taking these stories more seriously, offering "I chalked it up to a '50s–'60s era image of a boss chasing a secretary around the desk ... as if that's O.K. That's the egg on my face right now" (Kantor 2017). In this exchange, it's clear that Tarantino, while apologizing, is at another level excusing himself. Tarantino justified his inaction by saying that Hollywood had engaged in a kind of "Jim Crow–like system, that men have tolerated by not speaking up themselves and that other men in Hollywood should acknowledge this was wrong and 'do better by our sisters'" (Kantor 2017).[8]

The Comeback Kids and the Path to Redemption

Even if the path to redemption, or minimally, resuming one's career after allegations of sexual harassment and assault is not automatic, there is a growing list of people who had been accused of sexual harassment and assault who have been able to make a "comeback," that is, are able to work again after accusations have been leveled. One of the more insidious means of re-entry is to look at the ways that people like Harvey Weinstein were, at least before his conviction, allowed to attend some entertainment events. For example, there is a showcase for performers called Actors Hours, founded by Alexandra Laliberte. On October 23, 2019, at a place called Downtime Bar in the East Village of New York, Weinstein attended the performance of young comedians. When people in the crowd confronted him during the show's intermission, the protestors were removed from the venue. This event was preceded by another Actors Hours event Weinstein attended on September 30, 2019. In both instances, the performers called Weinstein out but apparently were criticized by the hosts for "souring the mood" (da Costa 2019). Though Laliberte later posted on apology on Facebook for not removing Weinstein, the idea that Weinstein himself would feel comfortable attending a public entertainment event does speak to the larger issue not only of a sense of entitlement, but also of how influence is wielded in the entertainment industry.[9]

In some cases, the timing of the entry back into the business seems as if it is designed to be a short hiatus but one that doesn't derail a career for any real length of time because the industry has somehow calculated that the alleged harasser's value is worth more than the cost of any possible negative publicity. For example, in December 2017, Max Landis, a Hollywood screenwriter and son of the famous director John Landis, was accused by a woman of sexual misconduct. After these allegations, he was not allowed to participate in a possible sequel of the film *Bright* (2017), but as of January 2019, he was once again working actively in Hollywood, having written a film called *Shadow in the Cloud* that will star the actress Chloe Grace Moretz, as well as another film, a sci-fi thriller called *Deeper*, that will star Idris Elba.

After these announcements were made about Landis' return to Hollywood, another woman wrote an anonymous post on the online website Medium that, while she had been interviewed by *The Hollywood Reporter* in 2017 with her own accusations as well as other women who accused Landis, the piece was not run because all of the women, including herself, were afraid to go public with their names (Zimmerman 2019b). She was moved to come forward publicly in light of the fact that Landis is again working in Hollywood, having made a seeming comeback. Landis' accuser then went on to a describe an incident in 2012 on a trip for work where he expressed his interest in her, and she had made it clear that she was not interested in him sexually. Landis grabbed her at one point and pushed her on a bed and held her down and tried to take off her clothes. After struggling with him and telling him no repeatedly, she pretended to pass out and he rolled over but slept next to her the whole night, and she had to drive home with him the next day on a 100-mile trip. She said that after that, while Landis did not dispute what happened, he said it was due to an "unfortunate drunken misunderstanding" (Zimmerman 2019b).

Landis, then, accused the young woman of overreacting and manipulated her into believing that she had simply misunderstood him, but the rumors kept growing of eight other women with whom he had behaved in a similar manner. She said she went over all of the events of that evening, which she remembers vividly, of how the encounter was non-consensual, including the fact that he pinned her down and that she had bruises on her body and she had kept saying no. She also realized that the other stories that were coming out about him sounded familiar to her own experience. While she didn't tell anyone about the incident at first, except for her very close friends, she felt like she needed to come forward after realizing people were trying to revive his career. She also realized that he had a pattern with women and if she didn't come forward, he was likely to repeat this pattern of sexual assault (Zimmerman 2019b).

In defending himself against these allegations, Landis offered that while he was likely guilty of a "boundaries violation," in his view he believes that the woman and he just had "different definitions of what a 'move' is, and in his case, he thinks of it as 'actively trying to begin a sexual activity with someone.... I just also I really feel like I have to stick to my guns on this if I'm not going to have to think about it all the time'" (Zimmerman 2019b). Countering this version of events, the woman responded by noting "Many of us have been in situations in which consent is murky. Sometimes, one person feels uncomfortable and doesn't quite know how to say 'no,' or someone is more drunk than they should be, or maybe the consent wasn't as enthusiastic as it should have been. But this felt very different; it wasn't a case of blurred lines or mixed signals. I was very clear in saying 'no,' if it hadn't been clear enough that I had run away twice and he had to chase me down, or that I was trying to push him off of me. I can't say for certain that he would have raped me if I hadn't pretended to lose consciousness, but I know that being forcibly held down and coerced by someone much larger than you when you're inebriated and stranded is not okay" (Zimmerman 2019b). Despite the woman coming forward, then, as well as several subsequent women with similar stories, it wasn't enough to dissuade those in Hollywood from continuing to work with Landis after a brief hiatus.

This pattern of individuals taking a hiatus and then returning after accusations have been made has begun to emerge as a form of backlash against the original accusers. This was the case of Aziz Ansari, who had been called out for his aggressive behavior on a date. His half-hour Netflix series *Master of None* (Netflix, 2015–2017) was considered for a third season despite the accusations against him. Defending Ansari, Cindy Holland, the head of original content for Netflix, said on a press tour that "we certainly would be happy to make another season of *Master of None* with Aziz" (Holloway 2018). Louis C.K., who had earlier admitted to masturbating in front of female comedians, was another entertainer who revived his career by doing additional shows after a period of time. Matthew Weiner, in addition, who had been accused of requiring a female writer for his acclaimed series *Mad Men* (AMC, 2007–2015) be seen naked by him, was also back at work with his $70 million show *The Romanoffs* (Amazon Prime Video, 2018). James Franco, who had also been accused of sexual misconduct by several women, was back at work as well on the HBO show, *The Deuce* (2017–2019). In the world of film, similarly, there has also been a return of people who had been accused of harassment and worse, with director Bryan Singer, for example, who had been accused of sexually assaulting a 17-year-old in 2003, being allowed to direct a remake of the film *Red Sonja* (Gilbert 2018a). And, finally, there is the case of Roman Polanski, who continues to make films and who won several prizes for his new *An Officer and a Spy* (2019).

More generally, those in the industry who were accused of sexual harassment have found that there are many paths back into the business, with only a short time elapsing from the time they are accused of misconduct to being re-integrated into the industry. For example, John Lasseter was a co-founder and animation "guru" of Disney's Pixar Animation Studio. In 2009, a young female graphic designer named Cassandra Smolcic went to work there and, after five years, left. From the first day of her job there, she was warned by others that Lasseter was sexist and lewd and, while she was there, she experienced sexually explicit behavior and was told to not be in a room alone with him. He was eventually let go after it was revealed that he had engaged in harassing behavior, including "grabbing, kissing and lurid comments" (Avrich 2018). Still, on January 9, 2018, it was revealed that Lassiter had been hired to run Skydance Media, a new film and TV production company, after a six-month hiatus. Reacting to this news, Smolcic noted, "It was pretty shocking to hear Skydance catapulted John back into such an important leadership role without anyone testing out his merit as a changed man or a 'reformed sexist' on the job first" (Fry Schultz 2019). This re-emergence, coming less than a year after being accused by many women of unwanted touching, raised the question of what it means for men who have been accused of fostering a sexist culture and harassment to be hired back into Hollywood.

Though Lasseter was hired back, it was not without controversy, which also speaks to the impact of #MeToo on Hollywood. After Skydance Media hired him for a head animation job, the actress Emma Thompson was moved to write a public letter, which was published in the *Los Angeles Times,* to Skydance chief executive and Oracle head David

Ellison, telling them why she was dropping out of a film project with them called *Luck*. As she wrote in her letter, "It feels very odd to me that you and your company would consider hiring someone with Mr. Lasseter's pattern of misconduct given the present climate…. If a man has been touching women inappropriately for decades, why would a woman want to work for him if the only reason he's not touching them inappropriately now is that it says in his contract that he must behave 'professionally?'" (Zeitchik 2019). She then went on to ask why a man who has made "women at his companies feel undervalued and disrespected for decades," should think that he will now show respect for them for any other reason than to do what he is "required to perform by his coach, his therapist and his employment agreement" (Zeitchik 2019).

This public rebuke of Lasseter, who had been responsible for such Disney megahits as *Frozen* (2013), *Toy Story* (1995), *Moana* (2016) and *Up* (2009), was a clear demonstration that some actresses had the power to speak out about what they perceived as Hollywood's quick acceptance of men who had been harassers back into the business. At the same time, however, the fact that Lasseter was hired back so quickly signaled that the studio was willing to absorb any negative publicity they might incur.[10] And, for Thompson, the larger question is not whether or when harassers like John Lasseter are allowed to return to Hollywood, but what to do for all the people who were harmed by him and whose own careers were ruined. As she concluded, "Much has been said about giving John Lasseter a 'second chance' … but he is presumably being paid millions of dollars to receive that second chance. How much money are the employees at Skydance being paid to give him that second chance?" (Zeitchik 2019).

Fighting Back: When the Accused End Up Accusing

In December 2018, Yael Stone, who starred in the television series *Orange Is the New Black* (Netflix, 2013–2020), made an allegation against the actor Geoffrey Rush, accusing him of harassing her sexually when they had worked together on a play, *The Diary of a Madman*, in 2010 and 2011. She told the *New York Times* that in her native Australia, where libel and defamation laws put the burden of proof on the accuser, she couldn't speak out about it there and that it was hard for #MeToo stories to emerge in her country for that reason. Rush then denied the allegations and by April 2019, he won a judgment against the Australian newspaper *Telegraph* for printing the accusations of another woman who made the same claims as Stone.

In another case, in September 2019, a court in Paris found that journalist Sandra Muller had defamed a television executive who she had accused of sexual harassment (Breeden 2019b). The French court ordered her to pay the executive, Eric Brion, 15,000 euros or $16,500 for damages as well as an additional $5,500 in legal fees, and she was also ordered to remove any Twitter posts that referred to him. In their ruling they noted that "[s]he surpassed the acceptable limits of freedom of expression, as her comments descended into a personal attack" (Breeden 2019b). Ms. Muller was an important figure in the #MeToo movement in France, having started the #BalanceTonPorc, or #ExposeYourPig, Twitter campaign in 2017 when she identified Mr. Brion in a series of Twitter posts. She wrote about her experiences with Mr. Brion, including during an event at a television festival in Cannes, when he told her: "You have big breasts. You are my type of woman. I will make you orgasm all night" (Breeden 2019). When she posted these comments, she then encouraged other women to also speak out against sexual harassment and ultimately, tens of thousands of women responded with their own stories, which in turn opened up a wider conversation in France on sexual harassment.

Several months after Ms. Muller's Twitter posts, the executive Mr. Brion went forward and sued her for defamation. In his suit, he didn't deny making the remarks but said she was wrong for viewing them as sexual harassment and that he was defamed when she tied his case to that of Harvey Weinstein. Brion claimed that he was drunk when he made the remarks and that he apologized soon after he said them. He also pointed out that they didn't work together and he didn't have any influence over her career. The court ended up agreeing with Brion's case and ruled that Ms. Muller didn't have a "factual basis" to make the accusation and that by using the term "pig," and linking Weinstein to Brion, he was subject to "social reprobation" (Breeden 2019).

The fact that the accuser was herself sued points to one of the more powerful forms of backlash that has emerged in the #MeToo era. Ms. Muller went from being called a "silence breaker" by *Time* magazine in 2017, and one of the women collectively named by *Time* as Person of the Year, to having to pay a large sum of money to the man she had accused of sexual harassment. Ms. Muller's case also reveals some of the more reactionary responses in other countries to the #MeToo movement, in particular in France where it has exposed a cultural divide, as evidenced by the public letter signed by Catherine Deneuve and a hundred other women against the #BalanceTonPorc movement in France. The accused man became, in the eyes of the French court system, the one who suffered harm, or as he described his predicament, "You can't destroy a man's life … just because one evening he spoke to you inappropriately, without going any further" (Breeden 2019b).

For those women who might choose to follow in her footsteps, however, and speak out against inappropriate behavior, the verdict in this case would have the chilling effect of women not wanting to come forward for fear that they would be sued for damages. If this is the legacy of the #MeToo movement, then the backlash that has occurred in the #MeToo era suggests that the struggle continues, for how to address sexual harassment and assault without re-victimizing those who have been victimized in the first place? It also suggests that while changes have occurred and will continue to occur, there is a need to simultaneously address the ways in which reactions to this movement may have the unintended consequence of causing more harm to the very victims who come forward in the first place.

Another issue raised with the backlash is that the industry itself needs to shift its gaze from focusing on the men who committed these acts to profiling the victims who were more vulnerable and less famous than the men who committed the sexual misconduct. There is a sense in much of the coverage of these men that the way they have fallen is tragic and that the loss of them from the entertainment industry is itself a tragedy. There are many interviews given by these men where the tone is often about how sorry the man is for the transgression and his desire to move on from it. *The Hollywood Reporter*, for example, has had sympathetic portraits of men such as Jeffrey Tambor and Charlie Rose and have focused on how difficult their life has been since the accusations came forward, and how Charlie Rose is "lonely" since he was forced to leave his job (Castillo 2020). In another article, on Bernardo Bertolucci, the director of *Last Tango in Paris* (1972), they highlighted his defense of Kevin Spacey, who has been accused of sexual misconduct, but left out the fact that Bertolucci himself had subjected the female star Maria Schneider to a rape scene without her consent in order to get a "natural" reaction from her.

For writers like Monica Castillo, media coverage from *The Hollywood Reporter* and other outlets that focus on the tragedy for the men who are losing their jobs, falls short in addressing the ways the actions of these men have affected the women they abused. As she noted, "The *Hollywood Reporter* has simply become the mirror of its industry, not the place to hold it accountable. Hollywood bills itself a progressive and enlightened haven even as it exploits and excludes women and people of color. There are ways to talk about the lives and careers of these men without making them martyrs. If *Hollywood Reporter* is any indication to the attitudes of the industry's upper echelons, then it looks like the promise to make Hollywood a better place for women may only be sweet nothings" (Castillo 2020). It is this attitude, of seeing the accused as victims themselves, that not only symbolizes the backlash that has occurred but more perniciously, threatens to upend the calls for change that the #MeToo movement has advocated for in the entertainment industry. That is why it is critical, moving forward, to recast the focus on the women, who are more often than not at the lower echelons of the Hollywood hierarchy, and whose own experiences threaten to be overlooked in the rush to forgive those who committed the misconduct in the first place.

4

The Role of Comedy
in Telling #MeToo Stories

Since October of 2018, when the #MeToo movement became part of the national conversation, its impact on stories in films and television could be felt. Existing stories were being re-written and new stories were being created to address the topic of sexual harassment and assault. Because Hollywood itself had been ground zero, so to speak, for many of the accusations of harassment, script writers were some of the first individuals to translate these events into the films and television series for which they were creating content. In both scripted and non-scripted television series and films, from dramas to documentaries, from comedies to reality shows, the issues raised in the #MeToo era began to appear with more frequency. Of all the genres, comedy has arguably dealt with the topic of sexual harassment with the most frequency. This may be due in part to the fact that comedy can be both revolutionary as well as reactionary when dealing with sensitive topics. According to television scholar Emily Nussbaum, other genres have a harder time grappling with the topic of trauma, since, as she observed, "Maybe there's been a hesitation to deal with this head-on in drama, because drama does, to some extent, at least, require sincerity, and sincerity can be uncomfortable in talking about trauma and assault" (New Yorker Radio Hour 2019).

While there has been a virtual deluge of episodes on all kinds of television genres that covered some aspect of #MeToo, it may be the case that the genre of comedy has yielded the most varied kinds of stories. Sometimes the actions of the characters are portrayed as uniformly bad, while in others the question remains as to whether the acts themselves were wrong or deserve a more nuanced treatment. It's helpful, then, to look at comedies that deal with the effects of harassment both directly, as well as those that offer a more reflective view of the question of responsibility in the wake of #MeToo. We will see how, as divergent as the stories are, streaming services, network channels and cable tele-vision all attempted to capitalize on the currency that stories about #MeToo could provide for their viewing audiences. Hearing how the performers, directors and writers understood the stories they were cre-ating in their comedies as a reflection of the #MeToo movement is also instructive and are included as well.

Streaming Services Pivot to #MeToo Storylines

One of the ways writers and showrunners of various series on streaming services such as Netflix and Apple TV+ were able to make their series feel current was to draw on the themes around sexual harass-ment and assault. This happened in series that had been on for some seasons, including Netflix's *GLOW* (2017–present), *BoJack Horseman* (2014–2020) and *Unbreakable Kimmy Schmidt* (2015–2019). In addi-tion, for new shows, such as Apple 1's *Morning Show*, the storyline was also re-envisioned to make sexual harassment charges a central aspect of the narrative. For example, episode 5 of Season 2 of *GLOW*, a show about female wrestlers who are trying to make it in a male-dominated field, confronted the issue of sexual harassment directly. In "Perverts Are People Too," the actress-wrestler Ruth Wilder (Alison Brie), is asked to attend a business meeting in a hotel restaurant with the TV network chief in order to drum up support for the wrestling show. When she gets to the hotel restaurant, however, she finds out that the meeting has been moved to the hotel room of the TV network chief himself. Though the conversation begins innocently enough, it quickly shifts as the execu-tive asks Ruth to show him how she does a headlock, and then shoves his face into her chest. He then goes to the bathroom to turn on his spa tub, and at that point, Ruth runs out of the hotel room. Because she rejected his sexual advances, the network chief retaliates by moving the show from its prime-time slot to a slot at 2 a.m., when the majority of the tele-vision audience is asleep.

Sharing their impetus for writing this episode, the female writers of *GLOW* described having experienced similar unwanted sexual advances during their time in Hollywood and wanting to write something cen-tered on that even before the stories about Weinstein were made public. Rachel Shukert, writer of the episode above, offered, "It really happened independently of [the current #MeToo and Time's Up movements], which I think speaks to how endemic this kind of thing is…. The exact specifics of Ruth's story were not 100 percent what happened to any-one in the room, but we are all women in the business and have all been in situations that have suddenly sort of become sexual in a way that we didn't foresee them becoming" (Miller 2018). It was partway through the scripting of the episode that the allegations against Weinstein were made public, so by the time they shot the episode it suddenly "felt more important than it had two or three weeks earlier" (Miller 2018).

In response to the revelations by famous actresses such as Selma Hayek, Ashley Judd and Gwyneth Paltrow about Weinstein and not wanting to exactly duplicate them in her script, Shukert changed some of the details of the episode so that it would look like it was more part of the storylines in *GLOW*. Still, there were some similarities that could not be downplayed, such as when they try to show how, from Ruth's perspective, she was questioning her own judgment over whether it was innocent or not when she is told to go to the executive's hotel room, and hoping that it is indeed innocent. And, like many actresses in Hollywood who are admonished to stretch themselves as actors and do things that might make them uncomfortable, the episode also shows Ruth being confused about the discomfort she was feeling at being called to his hotel room but also feeling like she should be a professional about it and go ahead anyway. The writers also wanted to show what it was like to have a bubbly demeanor and to have that turned into a tool against you, with Ruth questioning whether she was leading the TV executive on with her open and warm personality. The episode also underlines how Ruth's moves as a wrestler, something she worked so hard to become, had been used as a vehicle for the executive to make a pass at her, and Ruth ultimately feels complicit in somehow "leading him on."

When her former best friend Debbie (Betty Gilpin) makes her feel guilty for rebuffing the executive, because it resulted in having their show removed from the better time slot, the writers wanted to show how the accuser is often challenged for their response to the harassment. Debbie tells her that she should have given him the impression that she would have slept with him, but that she couldn't because she has a fiancé or she is having her period—and when Ruth tells her she isn't that kind of person, she says, "What, an actress? I mean this is how this business works, Ruth. Men try shit, you have to pretend you like it until you don't have to anymore.... Feminism has principles, life has compromises" (Miller 2018). For Shukert, this scene was pivotal because it spoke to the issue of how other people respond when you speak up about being sexually harassed. Many of the women who have made compromises essentially said that, as actresses, they had put up with this kind of treatment so other women should also. In these ways, the writers of *GLOW* were hoping to show both sides of the debate around the politics of speaking up in the #MeToo era as well as their own internal

GLOW (Netflix, 2017–19): Ruth (Alison Brie, right) is plied with drinks before a sexual assault. Also shown: Paul Fitzgerald (as Tom Grant, left) and Andrew Friedman (as Glen Klitnick).

dialogue when they both want to be angry about the harassment and, by contrast, also want to be somehow above it (Miller 2018).[1]

In 2018, Netflix offered not only *GLOW,* with episodes focused on #MeToo, but also two other comedies as well: the animated show *BoJack Horseman* (2014–2020) as well as *Unbreakable Kimmy Schmidt* (2015–2019). The main theme in the first four seasons of *BoJack Horseman* is how Hollywood caters to a rich has-been actor who is abusive to his co-workers. Will Arnett plays the voice of an anthropomorphic horse named BoJack Horseman, a washed up actor whose claim to fame was a 1990s sitcom. The show also stars his ghostwriter Diane Nguyen (Alison Brie), who is trying to help him write his autobiography, as well as his agent Princess Carolyn (Amy Sedaris). The main character BoJack Horseman is an anti-hero, and while he is meant to show how damaging male behavior is, the show's creator, Raphael Bob Waksberg, was frustrated that he was such a popular character with the audience (Salmon 2018).

In the fifth season, they show how dysfunctional this character is, while at the same time Waksberg underscores how he also is not so different from other actors in Hollywood. In this way, Waksberg allows BoJack to "embody" the very questions raised in the "MeToo" era (Salmon 2018). Season 5 is important because it shows how BoJack is trying to be a better person and agrees to star in the TV drama *Philbert* to help his ex-manager and ex-girlfriend Princess Carolyn. He ends up trying to stand up for his co-lead and love interest on the show, Gina, when she is asked to film a gratuitous nude scene. The showrunner Flip McVicker (voiced by Rami Malek) tells BoJack that Gina can remain clothed if BoJack gets naked instead. Bojack feels trapped because if he walks away from the show, he will end up making Gina suffer. For most of the season there is an assault on women, and this is one time where the male lead is objectified. In the first episode of the fifth season, "The Light Bulb Scene" (original release date September 14, 2018), BoJack tries to reason with Flip, but to no avail. Flip orders BoJack to take off his robe, and when he refuses, Flip tries to rip the robe off him, and they struggle as Flip continues to try to undress him.

Though the scene is clearly one where BoJack is attacked, this incident isn't mentioned again in the rest of the season's episodes. Instead, BoJack defends women who were abused and never connects it to the attempt by Flip to physically undress him. In the 11th episode of the 5th season, "The Showstopper" (original release date September 14, 2018), BoJack ends up physically attacking Gina himself, and it isn't explained why. And while the show continues to talk about sexual assaults that occurred against the women on the show, the assault against BoJack is never raised. In Season 5, episode 4, "BoJack the Feminist" (original release date September 14, 2018), one of the characters, Princess Carolyn, is a producer who employs Bojack. Another character, voiced by Angela Bassett, is a talent agent who continues to work with another sexual harasser because she wants to be successful. At the end of the episode, Gina (Stephanie Beatriz), one of the two new female characters, continues to work with BoJack after he tried to choke her while he was in a drug-induced state. Thus, while the show doesn't deal head on with the effects of assault on male victims, it does raise the issue of the pervasive sexism and harassment that exists in Hollywood for its female characters.

Another show that incorporated #MeToo into its storyline was the Netflix series *Unbreakable Kimmy Schmidt* (2015–2019). The show was created by comedienne Tina Fey and Robert Carlock, and starred Ellie Kemper as Kimmy Schmidt. The series is about a 29-year-old woman named Kimmy who is trying to adjust to life after having been held captive by a doomsday cult in Durnsville, Indiana. She and three other women had been held there by a reverend named Richard Wayne Gary Wayne (Jon Hamm) for over fifteen years. Throughout the series, Kimmy is often portrayed as a naïve sexual assault survivor who was kidnapped and spent half her life underground as a captive while being systematically raped by her captor. Because she didn't want her life to be defined by her captivity, she moves to New York City for a new start and becomes friends with her landlady Lillian Kaushtupper (Carol Kane) and rooms with a struggling actor named Titus Andromedon (Tituss Burgess). She also becomes a nanny for Jacqueline Voorhees (Jane Krakowski), a depressed, wealthy woman.

The series was an enormous critical success and has been nominated for 18 Primetime Emmy Awards, including four nominations for Outstanding Comedy Series. Though a comedy, the show's head producer, Robert Carlock, felt that it was a natural fit for the show to look at #MeToo by having Kimmy herself become a harasser, even if she would normally be the opposite of someone who might be perceived as a predator, as a humorous way to raise the issue of sexual harassment. In the first episode of the fourth season, "Kimmy Is… Little Girl, Big City!" (first aired May 30, 2018), Kimmy begins a new job at Giztoob, working in Human Resources at a tech startup. She is told to fire a young man but doesn't want to hurt his feelings and tries to make him feel good by complimenting him on his eyes and giving him a massage on his shoulders and telling him to relax. She then tells him that "[m]aybe our friendship belongs in the night hours" (Angelo 2018). She is fired after being accused of sexual harassment.

Unbreakable Kimmy Schmidt (**Netflix, 2015–19**): **Kimmy (Ellie Kemper) responds to years of abuse by determining not to be a victim, a strategy whose limits become apparent.**

Season 4 also has a darker storyline in the third episode, "Party Monster: Scratching the Surface" (first aired on May 30, 2018), that involves a fake documentary called *Party Monster: Scratching the Surface*. The creator of the documentary turns out to be Kimmy's captor, Richard Wayne Gary Wayne, and Kimmy is horrified when she sees she is portrayed as his long-suffering partner while he is in prison and that there is a man called Fran Dodd who runs an organization to help "fight back in the war on men." Dodd believes Richard Wayne Gary Wayne is unjustly imprisoned and states that "[m]asculinity is being criminalized in this country ... and I want something done did about it." When Kimmy tells him that Wayne is responsible for his problems, Dodd says it is women's fault for causing all the problems, and that "[s]ociety used to make sense! Nuclear families, straight marriages, white quarterbacks. That's the world the Reverend [Wayne] was trying to get back to. The bunker was a return to traditional values." Though billed as a comedy, *Unbreakable Kimmy Schmidt* takes on a serious strain in the culture that imagines someone like Harvey Weinstein is being framed. And though there were always themes about sexual assault on the show, because Kimmy has been assaulted for much of her life, in this fourth and final season the themes emerge most clearly (Wilkinson 2018).

Streaming channels such as Apple TV+, while newer than Netflix, also wanted to gain currency by drawing on the themes of sexual harassment and assault. To that end, in the fall of 2019, their newly launched series *The Morning Show* (Apple TV+, 2019–present) placed sexual harassment at the center of the storyline. *The Morning Show* was originally conceived in 2017 and was inspired by Brian Stelter's nonfiction book, *Top of the Morning: Inside the Cutthroat World of Morning TV* (Grand Central Publishing 2013), about *The Today Show* (1952–present), and centered on how Ann Curry, a morning anchor on the show, was ousted in a political battle with Matt Lauer, another co-host. Lauer was eventually fired after being accused of raping a co-worker, Brooke Nevils.[2] However, when the first episode aired, "In the Dark Night of the Soul It's Always 3:30 in the Morning" (first aired November 1, 2019), the story begins with Mitch Kessler (Steve Carrell), a well-liked morning news program anchor, being fired after he was charged with sexual misconduct.

While the original premise of *The Morning Show* wasn't focused on sexual harassment, the creators ended up revising the storyline to focus on sexual misconduct by the main male anchor and brought on Kerry Ehrin as a new showrunner (Blake 2019). As Ehrin commented about the new focus of the show, "That's it—like, the whole thing. That's what the subject is. There's no subject bigger or more important right now"

(Blake 2019). And, while *The Morning Show* isn't supposedly modeled on Lauer, there are a number of references to the allegations Lauer was charged with, including having a button under Mitch's desk to lock his dressing-room door so women couldn't get out. In addition, and unlike in real life, they hire an investigator on the show to look at the culture of *The Morning Show* to see how it could have fostered an environment that allowed Kessler to be sexually inappropriate with his female co-workers.

In the third episode of the first season, "Chaos Is the New Cocaine" (original release date November 1, 2019), Kessler is talking to a director (Martin Short), and they are bonding over the fact that they were both charged with sexual misconduct, though in the case of Short, he was charged with rape. Mitch tells the director that he wants to make a documentary about the #MeToo movement. He says that he thinks that there are two phases of the movement: in the first wave are men accused of doing things that were "really bad," while in the second wave, the one that he is in, are men accused of engaging in relatively minor forms of misconduct. He tells the director, "You are actually a predator ... and people are going to want you to own that" (Blake 2019). The director then challenges Kessler to ask what he was then, compared to him. Kessler then tells him that he knows that for one thing, he is sure that he is "not you" (Blake 2019).

The Morning Show (Apple TV, 2019): Celebrity news anchor Mitchell Kessler (Steve Carell) finds himself isolated, as his prototype Matt Lauer did.

While a lot of television series have used #MeToo storylines to remain current, *The Morning Show* offers it as a central story arc and also takes it further in order to interrogate some of the assumptions of the movement itself. It also tries to offer a more insider look at how some men are talking, using terms like "neo–McCarthyism," "woke Twitter," and "sexual agency," and they have spent a lot of time trying to portray what it feels like to be a man who is accused of sexual harassment. In addition to offering the accused's point of view, they have other characters voice sympathy for Kessler, as when his executive producer Chip Black (Mark Duplass) agrees with Kessler when he says he is being challenged because he is a straight white male. Black tells him, "The whole #MeToo movement is probably an overcorrection for centuries of bad behavior that more enlightened men like you and me had nothing to do with." The show does portray, however, the ways that men like Kessler can abuse the power they have in the workplace, especially in a place like a morning news show with its air of warmth and intimacy. It also shows how many men believe they are not abusive because they never acted like Harvey Weinstein. This is brought home in a scene where his former co-anchor, Alex Levy (Jennifer Aniston) tells him that she is angry with him for his behavior. In his defense, he tells her that he never "coerced anybody," and that it was the "fault" of Weinstein that everyone is now being called to account for sexually inappropriate behavior.[3]

Kessler's own sense of being aggrieved is highlighted throughout the show and by beginning with him being fired, *The Morning Show* allows plenty of time to explore how the characters respond to the charges and how they try to orchestrate their comeback. The show also explores how the women who come forward with their accusations are treated, and how women from different generations may respond to the issue differently. As we see later in the discussion of *Murphy Brown* (1988–1998; 2018) and *The Good Fight* (CBS All Access, 2017–present), some recent television shows also draw on the idea of how the older generation of women, rather than being allies, can sometimes end up tacitly accepting and even reinforcing the mistreatment of a younger generation of women.

Network Channels Confront #MeToo

Network channels also had their share of comedies which explored themes related to #MeToo, and the reboot of CBS's *Murphy Brown* was one of them. The original show was a situation comedy series created by Diane English that first aired on November 14, 1988. The show starred

Candice Bergen as main character Murphy Brown, an investigative journalist for a fictional television morning news magazine called *Murphy in
the Morning*. *Murphy Brown* originally ran from 1988 to 1998, and was
revived in 2018 by the same network, which was facing its own scandal with the exit of President and Chief Executive Officer Les Moonves,
who had been accused by multiple women of sexual harassment. At a
Television Critics Association press tour, executive producer Diane
English was asked about the allegations against Moonves as she was trying to promote the re-make of *Murphy Brown*. She told the audience,
"On behalf of everybody on our show, we take the allegations of sexual
misconduct very seriously" (Birnbaum 2018). She also told the group
that she supported the investigations into Moonves and that prior to
the publication of the allegations, they had written an episode based on
the #MeToo movement. Voicing her support for the movement, English
offered that "[i]t's a powerful movement, we wanted to do it justice, and
the episode title is '#MurphyToo'" (Birnbaum 2018).

In the third episode of Season 1 of the revived show (first aired
October 11, 2018) the staff of *Murphy in the Morning* are asked to sit
through a sexual harassment seminar. One of Murphy's colleagues,
Corky (Faith Ford), recounts all of the times she was confronted with
inappropriate behavior from men, saying, "I don't know any woman who
hasn't had an experience." Instead of agreeing with her, Murphy jok-
ingly tells her, "What guy would be stupid enough to try something with
me?," but then later tells her son Avery (Jake McDorman) that she had,
in fact, been sexually assaulted when she was younger. She then recalls
a traumatic incident that occurred when she was 19 years old, after she
received an award as a student reporter. She was sexually attacked by a
professor, but in the re-telling of the incident she at first tries to excuse
his behavior while questioning her own. As the episode unfolds, she
is hesitant at first to say much, telling her son, "something happened
… that's it," and then, "I put it in a drawer in the back of my brain and
moved on…. That's what we did in those days. It was a different time."
Later, Murphy talks to her friend Phyllis (Tyne Daly), a female character of the same age, and asks her for some advice about what to do about
her memory of the assault. Phyllis commiserates with her about what
it was like back then, telling her: "We flattered egos, laughed at lousy
jokes, and if something happened, we didn't talk about it. In those days,
it wasn't sexual harassment. It was a bad date."

As the episode unfolds, Murphy makes the decision to confront her
former professor. When she does, he responds by minimizing and dis-
missing her version of events, explicitly criticizing the #MeToo movement by saying it's just "women dredging up the past, pointing fingers,

ruining reputations." Realizing she won't get any closure by confronting her abuser, she instead retrieves the award she had won and which she left at his house the night of the assault. Symbolically, the professor had not only kept the award, but had put it on display in his office for all those years, as if his earlier conquest of her had somehow been a trophy. Though the episode was mixed in terms of its reviews, there was agreement that it had made a real attempt to see what it meant for older women to come to terms with events from the past that could be seen in a new light in the #MeToo era. Liz Shannon Miller offered this generational understanding of the episode, noting, "'#MurphyToo' is best read as what it is: an older generation of women not quite sure what to make of how things have changed. Bergen is 72 years old, English is 70, and when Murphy and Phyllis talk about 'the way things used to be,' it's an authentic moment. The ending gives her a sense of triumph, but the most powerful part of the story comes with Murphy's initial denial that comes with trying to survive" (Miller 2018).

Another show that made a conscious effort to explore the issue of sexual assault was *Brooklyn Nine-Nine* (Fox, 2013–2018, NBC, 2019–present). The show is a police procedural comedy created by Dan Goor and Michael Schur about a New York City police detective named Jake Peralta (Andy Samberg) who is goofy but talented and works out of a fictional Brooklyn 99th Precinct. He often clashes with his new commanding officer, Captain Raymond Holt (Andre Braugher), who is serious and stern compared to his more immature personality. Season 6, episode 8, "He Said, She Said" (first aired February 28, 2019), dealt with a sexual assault case. The detectives on the case, Amy Santiago (Melissa Fumero) and Rosa Diaz (Stephanie Beatriz), ended up disagreeing over the proceedings of the case, demonstrating that even among women, there is a diversity of opinions about issues around sexual assault. The case involved a finance "bro," Seth, who accused his co-worker, Keri, of hitting his penis with a golf club during a disagreement at work, which resulted in a penile fracture. Keri counter claimed that she was acting in self-defense. Keri was offered $2.5 million by the company to remain quiet about the incident, but Amy encouraged her to press charges anyway. Rosa, on the other hand, asks Amy to think about what it would mean for Keri to press charges, saying, "Let's just say, best case scenario you do find evidence [against Seth].... She's still going to have to go through a very public trial where they drag her name through the mud. Even if she wins, she loses. It's two steps forward, one step back." Amy counters by encouraging Keri to speak up, since it could inspire other women to do the same. The episode also offered a nuanced view of the tensions involved in pressing charges in cases of sexual harassment and assault. It is often not black and white how the harasser will present himself to others or even himself. Seth, the harasser, for example, believes he is a kind of proto-feminist, as when he says, "I'm the kind of guy who thinks Kathryn Bigelow should direct the next *Star Wars*—and I've said that to other men."

By offering different perspectives on the best course of action to take, the writers were trying to show that pressing charges can have repercussions for the woman who comes forward. Women have been challenged about whether they are telling the truth when they accuse someone of sexual harassment or assault, and can themselves lose their jobs. The episode also shows how hard it is to move forward legally with an assault case—when the detectives try to find evidence about an assault, they are often thwarted by the company's general counsel, who has already given the rest of the staff the same talking points to defend the accused. Even for those who do come forward, they still have to continue to live their lives, and to have gainful employment. This is demonstrated not only for the victim in the case, Keri, but as it turns out, Detective Santiago herself when she discloses that she had also been a victim of sexual harassment in her career as she moved through the police ranks to the title of detective. Her admission of this history of sexual harassment helps the audience understand that sexual harassment is not an isolated occurrence, but one that is frequently experienced by many women in the workplace. And, after discouraging Keri in the beginning from going forward with her claim, Detective Diaz ends up encouraging another colleague of Keri's to come forward, as the colleague reports her own experience having been assaulted. In this way, *Brooklyn Nine-Nine*'s writers wanted to show how all the characters learned from the incident about why it is important to speak up, rather than remain silent in the face of sexual harassment. In the end, Keri does go ahead and presses charges, and wins her case. She also ends up quitting her job, after she realizes that the whole company has "bros" who are similar to Seth and that she could find a better work environment elsewhere. Detective Diaz tells Amy, "Two steps forward, one step back is still one step forward."

Cable Television's Take on #MeToo

Cable television also brought forward storylines which drew on issues raised in the era of #MeToo. *Younger* (TV Land, 2015–present) stars Sutton Foster as Liza Miller, a 40-year-old divorcee who finds that she is too "old" to return to the field of publishing and ends up faking her identity to pose as a younger woman in order to get a job. The series has drawn on different figures in the literary and media worlds to create characters, including Edward L.L. Moore (Richard Masur), who is loosely based on George R.R. Martin, the author of *A Song of Ice and Fire*. Moore is portrayed as a key figure in keeping the publishing house afloat financially through the strength of his series. Moore's character, however, from the very beginning of the series, was drawn as a kind of lecherous figure, and Liza is exhorted by her bosses to keep him happy and to make sure he doesn't go to another publishing house. Though he was portrayed as a sexist character in earlier seasons, by the fifth season of *Younger* (2017), the writers decided to explore more critically his character's motivations and actions around the women he worked with. The creator of *Younger*, Darren Star, described their need to re-think their characters because the #MeToo movement had shifted the cultural conversation, or as he put it, "we had to think about what that cultural moment was saying about our characters ... and it was saying a lot" (Bennett 2018).

In the fifth season premiere, "#LizaToo" (first aired June 5, 2018), the editors of the company, Empirical, are dealing with an allegation that Moore had sexually harassed someone. One of the younger editors thinks that it is simply an effort to disparage Moore and most likely came from another publishing house. However, the editor and Liza's boss Charles (Peter Hermann) ask her point blank what her opinion is about publicizing Moore's next book. Diana (Miriam Shor), their marketing executive, also asks Liza whether she wants to say something and tells her they will support her, though it is clear she is doing this as a way to demonstrate solidarity when in reality she is anything but supportive of Liza. Liza looks around the room and tells them, "He's a flirty old man, but he never crossed the line, no." Based on this, the company decides to go ahead and make an announcement at the upcoming Comic-Con about Moore's newest book, but in the scene before the presentation, Moore again makes a sleazy comment to Liza. On the stage with Charles are several women who had formerly played Princess Pam Pam, a character in Moore's series, and they are all wearing the same skimpy costume that Moore envisioned for the female character he created as wearing. Liza was also forced to wear the same costume, and she finds out that all these other Pam Pams had also had the same experience of unwanted comments and sexual propositions from Moore. As Moore is then lowered onto the stage from above, the editor at Empirical makes the decision to postpone the announcement and publication of Moore's book until further notice. For Star, this represented a shift not only in their own thinking about the character of Moore, but also his perception that the larger culture had made a shift and that, as he noted, "we really had to take ourselves to task for how we approached a character like that" (Bennett 2018).

Another theme raised in "#LizaToo" deals with the ramifications of what had been a sub-theme in the show from the first season—the potential romance between Liza and her boss, Charles. The writers decided to deal with this issue directly, since it also had a different meaning within the context of the #MeToo movement, where relationships between bosses and their assistants are called into question because of the fundamental imbalance of power. During the season's finale, this issue comes up again, when a potential investor in the company raises it as a problem. The writers resolve the dilemma by having Charles assume a lower position so that he would no longer be the public face of the company. For the writers of *Younger*, this demotion was a way to show that the relationship between a boss and their employee has to be taken into account in a way that they had not dealt with in earlier seasons.

Jane the Virgin (CW, 2014–2019) was another cable channel show that intentionally focused on a theme from the #MeToo era, which, like *Murphy Brown*'s realization, has to do with how one retrospectively reads actions that occurred in the past. *Jane the Virgin* is a romantic "dramedy" and satirical telenovela (based loosely on a Venezuelan telenovela) that premiered on October 13, 2014, and was developed by Jennie Snyder Urman. The show stars Gina Rodriguez in the title role of Jane Villanueva, a twenty-three-year-old virgin who became pregnant after accidentally being artificially inseminated by her gynecologist. The show received a lot of critical praise, and was nominated for the Best Television Series—Musical or Comedy at the 72nd Golden Globe Awards, and Rodriguez won the award for Best Actress—Television Series Musical or Comedy. In Season 4, episode 11, "Chapter 75" (original air date March 2, 2018), Jane reflects on an affair she had with her advisor while she was in graduate school and thinks about the affair in a new light. At the time, the relationship was consensual but as she has gotten older, she begins to question the propriety of her advisor getting into this kind of relationship with a student. In handling it this way, the writers wanted to do more than simply write in a backstory of assault the character had never discussed before, or even add one to the present storyline. Instead, they wanted to take what had been a romantic storyline and show it from a new angle. As Urman explained, "The events that we showed were perfectly fine, but what if there was more to the story that you didn't know? ... So much of what you're realizing now is that there are patterns, and you're one of many, and people who transgress [do so] often usually. We wanted to figure out a way within Jane's world on how to engage with what's happening right now" (Turchiano 2018).

The way the writers ended up handling it was to have Jane see her former professor and lover Chavez kiss another student, which leads her to question whether going after younger female students had been a pattern in his life. In addition, at the time, Jane had ended up dropping him as her advisor so they could pursue a romantic relationship, which also makes Jane now question the choices she made and that he allowed her to make. Jane questions whether she would have made the same choice if she knew that this was a pattern for him of dating younger women who were his students. In so doing, she begins to look back at the relationship as one where there were unequal power dynamics. As Urman explained, "We wanted to look at the gray area, so that Jane could be working it out.... Even though it was fine [at the time], it doesn't feel as fine in retrospect. We wanted to see how this moment has changed the way people look at stories and the way Jane looks at something that happened post–the Me Too movement" (Turchiano 2018).

Another way of thinking about the impact of the #MeToo era on comedy is to contextualize it within the experience of actors and actresses who had their own #MeToo experiences and then see how they used it in subsequent storylines. The cable channel HBO offers several comedies that have raised issues around #MeToo. Ilana Glazer, for example, the co-creator and co-star of HBO's *Broad City* (2014–2019), was herself a victim of sexual harassment and spoke openly about her experiences in middle school and high school, offering, "I've been sexually harassed countless times. In middle school, in high school—by more teachers than students! At work as a waitress, at work as a showrunner! Same same same—I was a woman in both places. I was sexually harassed by a creepya-- doctor just last year and filed a complaint with NYC" (Romano 2017). When asked whether she was sure this happened to her, she then says, "Hm. Okay yeah lemme think a sec—YEAH I'M F-ING SURE," and said that while being harassed was a constant, being able to do something about it is much rarer (Romano 2017). She then went on to thank other actresses like Viola Davis and Tracey Lysette from *Transparent* (Amazon Prime Video, 2014–2019) for being so open about their experiences, which allowed her to come out and speak about her own.

Broad City is a situation comedy, based on an independent web series with the same name, that stars Ilana Glazer and Abbi Jacobson. The show draws on their own real-life friendship and how they are both struggling to make it in New York City. The show originally aired on Comedy Central on January 22, 2014, and has received critical acclaim for its originality. In Season 4, episode 8, "Housesitting" (original air date November 15, 2017), Glazer confronts the issue of sexual harassment directly and in a way that suggests that comedy is able to make the topic more nuanced and counterintuitive than it may at first seem. Roomates Ilana and Abbi are back from Florida, and Ilana is also back together with her boyfriend Lincoln. She is housesitting for her friend Oliver's mom in what can be described as a dream house for a New Yorker, one where she can run around and enjoy the expensive furniture, use the bidet and do months' worth of laundry. Abbi decides to do "Bumble," the online dating match-up app, and finds that her handsome, old high school teacher Richard (Mike Birbiglia) is on the site, so she "swipes" him and invites him to meet her and catch up on old times.

When Richard arrives, they begin to talk and he admits that he had found her attractive when she was in high school, at which point Abbi abruptly makes an exit. Ilana, when hearing that Abbi's old teacher had found her attractive, instead of being "creeped out" by that, instead tells Abbi that Richard had definitely masturbated to her back then, and that he should be considered laudable because he never actually tried anything with her. When Abbi and Richard get together again and eventually begin to make out, Abbi is uncomfortable with the fact that Richard wants to engage in student-teacher role play with her as part of their sexual encounter. He tells her that she can pretend to be seventeen and even pulls back her skin so she would look younger. When she tells him she is uncomfortable, he indignantly tells her, "We were role-playing!" At that point, their conversation is cut short as a house fire alarm goes off. While they are standing outside, Abbi asks Richard directly if he masturbated while thinking of his students, and he tells her, "All teachers do.... You kind of have to." In this way, *Broad City* both raises and deals more ambiguously with the problem of sexual harassment, not by demonizing the harasser but by leaving it open as to whether this is problematic or merely human.

Film and Television Comedies Featuring Abusive Women

One of the counterintuitive responses to the #MeToo movement in television series and films was that many of them portrayed women as abusers, rather than victims. Although in real life the actual number of women accused of sexual harassment and assault was very small compared to men, in film comedies like the 2019 *Late Night* or the television series *Great News* (NBC, 2017–2018) women were portrayed as engaging in inappropriate conduct. *Great News*, for example, is a comedy about a television news show that had a storyline about the repercussions of sexual harassment. The premise of the series is that it is set in the world of television news and explores the life of a female news producer who has to deal with a new intern, who also happens to be her mother. An episode focused on Tina Fey (who is also the series executive producer) who guest stars as a female executive who harasses people in the office to get a "golden parachute" in order to be able to retire from the job with a high payout. The storyline mirrored the experiences of real-life media figures like Fox's Bill O'Reilly and Roger Ailes, both of whom received huge payouts in order to leave their jobs in the wake of sexual harassment allegations. In the case of Fey, the storyline was focused more on the issue of her exhaustion at "having it all," as a woman with a high-powered job trying to balance her work and family life responsibilities. The show's creator, Tracey Wigfield, wanted to demonstrate that there was still a lot of victim-blaming going on in these cases and that by reversing the genders, she would be able to show how unfair it was (Friedlander 2018). As Wigfield offered, "Even for me, as a woman, I feel like those questions would kind of pop up, so the gender reversal would illuminate how absurd it was to blame women for bringing on this kind of harassment from their harassers" (Friedlander 2018).

Another portrayal of a female boss who engages in discriminatory acts against her underlings was seen in the theatrically released comedy *Late Night*, starring Mindy Kaling (who also wrote the comedy) as a new writer on a late-night show and Emma Thompson, who portrays Kaling's boss and who is also the host of the show. The film, which looks at the ways that women are discriminated against in TV writers' rooms, shows how Thompson's character, Katherine Newbury, is forced to own up to the fact that she has been having an affair with one of the writers on her show, Charlie (Hugh Dancy). The film explores more generally the way in which TV writers' rooms are routinely filled with white males, even if there is a female boss at the helm. Through the plotline of sleeping with an underling, the film does evoke other recent harassment cases, including that of the late-night host David Letterman, where he had been blackmailed by someone who accused him of sleeping with people on his staff for decades. In *Late Night*, Charlie is not simply an underling who is being exploited, but is portrayed as being manipulative for his own gain and encouraging Katherine to sleep with him when her husband is diagnosed with Parkinson's disease. Despite his manipulation, Katherine explains that she is just as bad as a male boss who has slept with an underling because she has the power to fire Charlie and to retaliate against him if she wants to. In this way, Kaling reverses the script of the woman who is being retaliated against to show that women in power are capable of exploitation in the same way that men are. Commenting on this role reversal, film critic Inkoo Kang has noted that "Kaling's script feels like a complication of the blanket rules that came out of #MeToo rather than an exemplification of them. It's understandable that female creatives would rather dwell on what women are capable of rather than the age-old reality of male domination. But the movement has been so exhilarating in its demand that we pay attention to the nuances of social dynamics that reductive gender flips, like *Late Night*'s, feel like a step backward" (Kang 2019).

Yet another show which focuses on a negative female character within a #MeToo context is the HBO award-winning series *Veep* (2012–2019), starring Julia Louis-Dreyfus as Selina Meyer, a foul-mouthed politician who is an equal opportunity offender of both men and women. In Season 7, the show directly confronts issues raised by the #MeToo movement, with Meyer questioning why it was that if she, as a woman, had to put up with men's bad behavior all these years, couldn't women in politics today do the same? Selina is running for president and using a #MeToo moment to bring down another candidate, Tom Jane (Hugh Laurie), who had been her boyfriend in previous seasons. She challenges his chief of staff (Rhea Seehorn) to accuse Tom of sexually harassing her, despite the fact that she is willingly involved in a romantic relationship with him. When the chief of staff goes ahead and accuses him, it ends up ending his presidential campaign and his marriage, both at the same time. This episode of *Veep* drew on themes from the #MeToo era to illustrate not only how women in power can be just as unethical as men in power, but that they have additional tools to draw on by virtue of their being women.

Summary: Too Much Freedom?

In many ways, the comedies that drew on #MeToo have invited audiences to similarly reflect on the changing assumptions and attitudes held in earlier generations about what was considered appropriate and inappropriate behavior. There is a sense, in watching episodes on many comedy television series that had previously not dealt with the issue of sexual harassment, that the writers are trying to take some responsibility for making it part of the cultural conversation. What all these shows and recent films have in common is that they are grappling, in different ways, with the meaning of #MeToo in the context of their stories and character portraits. These efforts have been part of a larger cultural conversation going on in Hollywood, which is trying to advocate for larger changes in the industry in the wake of #MeToo.

At a 2019 AMC (American Multi-Cinema) Summit, for example, there was a discussion about the way comedy has changed in the #MeToo era. The moderator Jill Kargman, the creator of the Bravo series *Odd Mom Out* (2015–2020), asked the panelists, who were actors, writers and showrunners, what the challenges were of trying to be funny in the #MeToo era (Edelstein 2019). This was part of a larger question of asking what the shifting limits were in comedy in a new age where there are a variety of new features, including trying to figure out what is and isn't appropriate and the relatively new freedoms that have come about as a result of the increase in content options in the streaming era. Some of the panelists said that there should be no limits placed on what is considered appropriate or inappropriate and that writers' should have the freedom to imagine whatever they want. Richard Kind, for example, who stars in the IFC mockumentary series *Documentary Now!* (2015–present), defended this view, noting that "I believe that in the writers' room—some may call it locker room humor—but in the writers room, everything is safe. There may be mistakes made, prejudices that come out of people's mouths, they are tempered by each other, and then what comes out of the writers' room is something that has been gone over by the writers and then makes it to America. Some really horrible stuff may lead to real brilliance" (Edelstein 2019). Other panelists, including Sally Woodward Gentle, the executive producer of *Killing Eve* (BBC America, 2018–present), felt that this kind of license should be allowed for writers' rooms which are female-dominated. Her perspective was that while men were allowed the freedom to be uncensored, women were expected to have more "decorum," but that this should not be the case because "[Women are] filthy and f*cked up and inappropriate. We have men in our room, but in the end it's these women giving amazing female actors voices. We've never self-censored" (Edelstein 2019). The panel, more generally, argued against the idea that the #MeToo era has led to censorship in comedy and instead pointed out that there are more opportunities than ever before. This includes the idea that the industry is now looking for more diverse perspectives, including those from different cultural groups, the disabled and the LGBTQ community. The pressure, if anything, is to be more ambitious, innovative and creative, and now more space has finally opened up for different groups to produce their work.

If the writers and creators of comedy are feeling less censored, rather than more, in the wake of the #MeToo era, another conclusion is that there is more space for stories that embody some of the most important messages from the movement. For example, the romantic comedy genre has been viewed in American film history as oftentimes regressive in its views about male-female relations, and the genre itself has been decimated by the film industry since its relatively more popular period in the 1990s, when films like *Sleepless in Seattle* (1993) and *You've Got Mail* (1998) were made. Since that time, however, there have been fewer romantic comedies made, and writers like Andrew Romano have even declared "The Romantic Comedy Is Dead" (Smith 2018). In 2018, however, the film *To All the Boys I've Loved Before* came out, based on the 2014 young adult novel by Jenny Han (Simon & Schuster), which in turn led critics to predict that it would lead to a "rom com revival" (Smith 2018). The story, directed by Susan Johnson, follows the traditional storyline of a romantic comedy where there is a "meet cute," then a complication and then a resolution, and centers on a 16-year-old named Lara Jean (Lana Condor) who writes love letters to all the boys she's loved. She stores them in a hat box without ever intending to send them, but her little sister Kitty (Anna Cathcart) mails them. For our purposes, it is instructive because the heroine in *To All the Boys I've Loved Before* behaves differently than other heroines in traditional romantic comedies.

In conventional romantic comedies, there is usually the construction of the female as essentially passive and waiting for the clueless male to realize they are in love with her. In this scenario, the boy is the one who gets to decide when the relationship begins. Though initially it would seem that Lara Jean is passive and doesn't have agency (because it was her sister who mailed the letters), she takes charge of her situation and is the one who decides when, and with whom, to have a relationship. At one point, she draws up a fake relationship contract with Peter Kavinsky (Noah Centineo), one of the most popular boys in the school, in order to throw Josh (Israel Broussard), another boy, off the trail that she was in love with him. Peter would also benefit from this fake relationship because he wants to make his ex-girlfriend (Emilija Baranac) jealous. One of the #MeToo themes underlying the film is that Peter is shown to be respectful of Lara Jean and aware of her sexual boundaries. This is in direct contrast to earlier films such as *The Breakfast Club* (1985), where the character John Bender (Judd Nelson), for example, hides underneath a table where Molly Ringwald's character is sitting, with the implication being that he is looking up her skirt and touching her. In another scene from *The Breakfast Club*, one of the young women in the film gets drunk at a party and passes out. The hero of the film Jake Ryan then says he could "violate her ten different ways if [he] wanted to" (Smith 2018). By respecting her sexual boundaries, Peter is directly countering these earlier portraits of young men as inherently predatory, which the audience is supposed to read as being "romantic."

Another way Lara Jean transcends conventional romantic comedy heroines is by being the one who has agency and declares her love, rather than the other way around. This subverts the declaration of love from the male, which characterizes earlier romantic comedies and which is usually the end product of a chase where the woman is essentially stalked by the male. The film upends these traditional assumptions about men and women in romantic comedies by having the young woman declare her feelings, but in a way that is respectful of the young man as well. A final theme in the film which resonates with the #MeToo era is Lara Jean defending herself in the face of being "slut shamed." At the end of the film, Peter's ex-girlfriend, Gen, tries to slut shame Lara Jean by posting a "sex tape" on Instagram, which shows Lara Jean and Peter in a hot tub. Though Peter wants to confront Gen himself, Lara Jean tells him, "This is a fight I have to handle myself," and she tells Gen, "It's bad enough if a guy were to do this, but the fact that a girl did? It's despicable" (Smith 2018). In these and other ways, the female heroine gets to decide her own fate and fight her own battles, with the young man himself encouraging her in the end to be clear and tell him what she wants. This is evidenced in the last scenes of the film, where Peter sees that Lara Jean is holding a letter and, though he teases her at first by grabbing the letter, when he realizes it's a love letter for him, he tells her that she must give it to him, rather than him grabbing it from her. This puts her in the position of being active, rather than passive, in her declaration of feelings for him. Carolyn Smith (2018), noting the revolutionary nature of this transformation of the romantic comedy, offered: "In rewriting the roles of the romantic comedy heroine and hero, Jenny Han and Susan Johnson present viewers with a romantic comedy that updates the genre's gender dynamics for the '#MeToo' era. It's about time."

If there are some ways to summarize these shifts in comedies, it is in seeing how they have challenged their audiences to realize that earlier jokes about women are no longer taken at face value and are challenging us to see what we consider funny without giving up our own humanity. The issues that have made women in the larger culture uncomfortable and have translated into speaking up about harassment are portrayed in these shows as problematic as well. Sometimes this discomfort is expressed through satire, while in other ways there is an attempt to make women in the audience feel empowered by showing these experiences as being worthy of speaking up about. This doesn't mean that there aren't jokes that will be made which may highlight the issues, or that comedians who have themselves committed sexual harassment and assault won't themselves be joked about in highlighting the problem of harassment. The larger point is that comedy has served as a vehicle to process the larger culture's response to the issues raised in the wake of the #MeToo movement, or as Molly Ivins has written, "Satire is traditionally the weapon of the powerless against the powerful" (Jasheway 2018). As we will see in the next chapter on dramas and documentaries in the #MeToo era, these voices emerging in comedy can themselves be viewed as a kind of cultural weapon in the effort to address the sexual harassment problem of which the #MeToo movement raised awareness.

5

Dramas and Documentaries in the Era of #MeToo

Dramas

By the 2017–2018 television season, dramatic stories began to appear that also drew on issues raised in the wake of the #MeToo movement. Though there were the usual constraints of programming in terms of production, in retrospect, the timeline was relatively short, and some have speculated that this was because the issue had been around for a long time in the entertainment industry. As we will see, it was not only comedies and "dramedies" that were able to draw on #MeToo themes, but dramas and documentaries also mined this terrain to create powerful entertainment around this subject.

ABC's *The Good Doctor* (2017–present), for example, about an autistic doctor, Shaun Murphy (Freddie Highmore), intentionally wrote in a storyline around a woman who is sexually harassed by a superior at work. The creator of the show, David Shore, felt it was important to include this kind of storyline because "It's a real issue and it's been a real issue for a long time ... in its most egregious form and in its subtle form ... when the idea of that came up in the writers' room, it seemed important to do it" (Friedlander 2018). In Episode 10 of the first season, "Sacrifice" (original air date December 4, 2017), Dr. Claire Brown (Antonia Thomas), who works with Murphy, plays a younger doctor who is sexually harassed by an older doctor who is her supervisor. The writers chose this character because they thought she would be the most vulnerable as a young, attractive intern. While the episode was aired in their midseason finale, the writers also felt that it would be important to show that it wasn't just an isolated incident, but would continue for the remainder of the season because in real life, harassers usually have a pattern of harassment, and therefore it often isn't a one-time event. In the same vein, they also didn't want to make it an isolated story but wanted it to play out over several episodes.[1] Finally, the writers wanted to portray sexual harassment as something that is part of a broader phenomenon in the medical community, so they intimate that it happened to another character, Allegra Aoki (Tamlyn Tomita), who, unlike Claire, has a more powerful status as the chairperson and VP of the foundation in control of the hospital in which they work.

Another dramatic hospital show that carried a storyline influenced by #MeToo is the enormously popular *Grey's Anatomy* (ABC, 2005–present), created by the showrunner Shonda Rhimes. The series focuses on the lives of the surgical residents, interns and attending doctors in a hospital in Seattle and portrays how difficult it is for these surgeons to try to have a personal life while they are also working. It has been enormously successful throughout its run and is one of the top three dramas in all of broadcast television in the United States as well as being the longest running primetime scripted series for ABC and the longest running medical drama on American television.[2] In its sixth season (2009–2010), the writers introduced a character named Jackson Avery (Jesse Williams), the grandson of Harper Avery (Chelcie Ross), a founding partner and head of the foundation that helps to support the hospital. In Season 14, Harper Avery dies and Meredith Grey (Ellen Pompeo) is given the prestigious Harper Avery award, named after the founding patriarch.

In Season 14, episode 20, "Judgement Day" (originally aired April 19, 2018), the writers of *Grey's Anatomy* decide to make the central storyline revolve around sexual harassment. Jackson finds out that a woman has been barred from working with the hospital because of a mysterious agreement she had signed earlier, and the reason given by his mother Catherine Avery (Debbie Allen) seems implausible. After doing more digging, Jackson finds out that the woman, along with twelve other women, had been the victims of sexual harassment at the hands of Harper Avery and had all signed nondisclosure agreements to silence them from sharing what happened. In a direct reflection of the types of accusations leveled against Harvey Weinstein, the writers of *Grey's Anatomy* took a page from the headlines and used it to explore the ways company executives enable the bad behavior of their bosses and how companies try to shield themselves from any harm through having these kinds of nondisclosure agreements. In the storyline, though Catherine tries to stop Harper from committing these abuses, she nevertheless ends up settling with the women, which costs the foundation a tremendous amount of lost revenue. Jackson challenges the cover-up and shows horror at how these women's voices had been silenced through the nondisclosure agreements.

Catherine tries to defend herself to her son by saying "30 years ago, getting harassed at work, getting groped, wasn't something we couldn't protest. It was something we had to take with our morning coffee. It was part of the job." She then went on to describe how women could lose their job back then if they complained, and that one of the reasons she paid the women off was because she was afraid Harper would just fire the women and would challenge their stories and ruin their reputation. She then tells her son that she is ashamed of what Harper did and that because of this, and the financial implications of it, the hospital could also be ruined. When Meredith learns about Harper Avery's behavior, she returns the award. Catherine then offers to step down from her position as head of the foundation if it will allow the hospitals to continue to run, but her son stops her from doing so because he doesn't think she should have to take the fall for the actions of his grandfather. The fact that she is a black woman and Harper Avery is a white man also amplifies the racial politics on the show when Jackson tells Catherine that she shouldn't have to step down as the head of the foundation, and "There's no way a powerful black woman is going to take the fall for a rich old white man who couldn't keep his hands to himself." Instead of resigning, then, Catherine takes the name of Catherine Fox, her maiden name, and the Harper Avery Foundation is dissolved. A new foundation called the Catherine Fox Foundation, dedicated to healing the wounds caused by the sexual harassment of her father, is created in its stead. The new foundation helps the women who had been abused by Harper Avery through retraining and rehiring them to work at the hospital.

Though some women have applauded the storyline for raising issues around #MeToo, other fans had more trouble with the way in which the plot played out.[3] For these viewers, the notion that the woman who had protected her father would still be able to head a foundation is problematic, since the women who had suffered themselves at the hands of her father never got the opportunity to win the award themselves.[4] In these ways, perhaps unintentionally, the series ended up mirroring what oftentimes happens in real life, where the people who were the "enablers" of the harassers themselves don't suffer any repercussions.

Legal Dramas and the Fight Over #MeToo: The Case of The Good Fight

Legal dramas are another genre where the #MeToo movement resonated, most likely because these dramas dealt with the legal implications of harassment and assault and thus provided a fertile ground for thinking about how to address these private acts in a public way. One recent series with storylines that draw on themes in the #MeToo era is the All Access television series *The Good Fight* (2017–present). The main character Diane Lockhart (Christine Baranski) had earlier worked at a firm in *The Good Wife* (CBS, 2009–2016) and had her savings wiped out due to an enormous financial scandal. She and another young lawyer named Maia Rindell (Rose Leslie) end up joining another prestigious African American legal firm (Lockhart is white) that employs a former employee of Diane's, Lucca Quinn (Cush Jumbo). The series is characterized by stories that are "ripped from the headlines," and explores a variety of topical issues with commentary on the social and political meanings of these cases.

Robert King and Michelle King, the creators of *The Good Wife*, made this spinoff and devoted several episodes of their second season to issues around sexual misconduct and harassment.[5] For Michelle King, these stories were compelling, and she noted that "it was really on any given day what was grabbing us and causing all of us to argue in the writers' room.... If people can't let a subject go, and [are] having disagreements about it, that's a good indication" (Friedlander 2018). Robert King, her partner, also pointed out that these kinds of stories are important because they tap into how social media has an impact on the larger society and challenges writers to think about the disagreements people are having around what constitutes inappropriate behavior. He said, "Everyone in the writers' room ... has these massive arguments about the new rules and maybe we should put new rules in place. We honor that by exploring how we disagree on both sides" (Friedlander 2018).

In the 11th episode of the second season, "Day 478" (original air date, May 13, 2018), the plot revolves around a client the firm has taken on, a photographer who lost his job because he was named on a website called "Assholes to Avoid" after he was accused of sexual misconduct that happened while on a date with a woman. While the photographer and the woman agreed on what happened when they were together, they had different interpretations of why it happened. In her view, she was pressured into having sex with the photographer, while he maintained that it was based on mutual consent. Two female lawyers, Liz Reddick (Audra McDonald) and Maia Rindell (Rose Leslie), represent the man suing the woman who posted the account of their date, and the lawsuit eventually grows into a larger suit against the whole website as more men who are also named on the site begin to lose their jobs as well. One partner, Adrian Boseman (Delroy Lindo), argues that "Maybe #MeToo has gone too far.... I think good causes start out being good and end up becoming mobs." Liz asks him whether he would make the same argument about Black Lives Matter. She then tells him that "[w]omen join together and all of a sudden men all over the world are worried about mobs, or witch hunts, but you don't have the same worry about Black Lives Matter ruining white people's reputations."

At a later point in the episode, Diane Lockhart (Christine Baranski) sits in on the deposition of Gretchen (Zoe Winters), the young woman who created the website. Gretchen asks Diane exactly what her feminist credentials are. When Diane tells her that all of the women in the room are good feminists and that they agree with her, Gretchen tells Diane, "We don't need you. This is a young woman's fight." At a later point in the episode, Diane responds to Gretchen's critique and tells her, "Women aren't just one thing, and you don't get to determine what we are." Explaining how this episode was conceived, Michelle King noted that it drew on the Babe.net story about Aziz Ansari, a comedian accused by a young woman, Katie Way, of sexual coercion that happened while on a date with him. The episode also drew on a media fight that subsequently occurred between the MSNBC host Ashleigh Banfield and Katie Way, with Banfield, as an older woman, questioning the way Way had written the story. King offered that similar arguments were happening among the writers on the show. In her view, while it is possible to show a clear example of a sexual assault, it is in the gray areas where it is more interesting for the writers to portray the complexities involved in understanding why something is wrong (McHenry 2018).[6]

In Episode 5 of Season 2, "Day 436" (original air date, April 1, 2018), the firm defends a broadcast network that was going to do a show on a male movie star who had been accused of sexual assault. At the same time, the previously mentioned Adrian Boseman (Delroy Lindo) explores his own motivations about whether he had given his ex-wife Liz (Audra McDonald) preferential treatment when he was a lecturer at her law school. Michelle King wanted to explore the more complex ways that preferential treatment, or putting down of others' ambitions based on gender and sexuality, can also be understood as misconduct in the workplace. By showing it emanating from Adrian Boseman, one of the characters who is admired on the show, King wanted to demonstrate that even the most beloved characters need to explore their motivations in the era of #MeToo. As Robert King also observed, "We live in a world where we are dealing with female and male writers so closely in the writers' room…. You always go through, *Did you shut somebody down too quickly? Did you prevent somebody from blossoming into a better writer by shutting them down?* I think the show is best in my mind when it points out the subtle biases of characters" (McHenry 2018).

In the third episode of Season 2, "Day 422" (original air date, March

18, 2018), there is another storyline that deals with sexual assault and that mirrors the *Bachelor in Paradise* (ABC, 2014–present) reality show scandal. The law firm takes on a case of a scandal that took place on the set of a reality show called *Chicago Penthouse*. The producer shut down filming because a contestant named Melanie (Isabella Farrell) had sex with another contestant but was too drunk at the time to give consent. Melanie goes on to sue the network because she felt she wasn't protected by them. The network denies responsibility for the assault, but then it is discovered by another lawyer, Marissa Gold (Sarah Steele), that there is footage showing that the producer had taken Melanie's unconscious body to the hot tub where the assault took place. Michelle King explained that the motivation for doing this episode was the idea of bringing forward the issue of consent and how it plays out on reality television.[7] They also wanted to explore how reality shows trick the viewers into believing the actions of the characters are somehow "real."

When the Abused Becomes the Abuser

In some of the dramas that emerged during this period, the idea that a woman or man who has been abused can in turn become an abuser is a theme that is explored in dramatic form. As we will see when exploring horror films during this era, the theme of trauma as a legacy of sexual harassment and assault is one that is used to explain a character's motivation. And, like the idea of a woman who becomes an abuser, as was seen in comedies like *Late Night* and *Great News,* in dramas there are also counter examples of strong female characters who are manipulative and aggressive. This can be found in the second season of the HBO drama *Big Little Lies* (2017–2019). Though the first season focused on the question of domestic violence and assault committed by Perry (Alexander Skarsgard), a man who had been abusing his wife, Celeste (Nicole Kidman), for years, the second season focused on a female abuser. The plot twist was that the abuser, Mary Louise (Meryl Streep), was the mother of Perry, who had been murdered in the first season. The show portrays Mary Louise as challenging Celeste's accusation that her son was a domestic assaulter and rapist.

While the question of domestic violence hasn't been front and center in the discussions around #MeToo, it has also been true that many similar stories are able to illustrate some of the dynamics involved in keeping silent after assaults, as well as the lingering trauma associated with sexual assault. In *Big Little Lies*, the wife is by no means uniform in her assessment of her husband, despite his violence toward her. In addition, the ways that violence can also have lingering effects long after the attacks took place is also explored, including having the victim commit acts of assault as well. This is portrayed by Celeste acting violent herself, punching her son, after she tries to break up a fight between her two children. In addition, the show also tries to be more nuanced in its portrayal of Jane Chapman (Shailene Woodley), the character who was raped, and in this way challenges the idea that the victim needs to be "perfect" in order to be believed (Kang 2019). Speaking about the ways that *Big Little Lies* connected with the #MeToo movement, Kidman, who helped bring the series to fruition, described how it became a word-of-mouth hit because it resonated with the themes that were being raised in the #MeToo era (Levine 2018). As she noted, "We started our production well before #MeToo. And then it was percolating a bit when *Big Little Lies* first came out. That is why for that role of Celeste, there was a bigger response than anything I had done before" (Levine 2018). In her view, her character leads a privileged life, and she is portrayed as staying in an abusive relationship for the sake of the kids as well as to be able to continue to have that life. This thinking about why women stay in domestic violence relationships in turn "started that whole conversation about hidden violence, behind closed doors" (Levine 2018).

Documentaries and Docudramas in the #MeToo Era

In many ways, the most compelling stories that emerged in the wake of #MeToo were ones that profiled real-world victims and perpetrators of sexual harassment and assault. Several shows, including both documentaries and docudramas, emerged that drew on and profiled the actions of real-world perpetrators. One documentary that emerged was the 2018 film called *Rocking the Couch*, written and directed by Minh Collins and produced by Andrea Evans. The documentary focuses on the early 1990s but situates the viewer in even earlier accounts of sexual harassment and assault from the beginning of the entertainment industry in Hollywood.[8] The documentary then goes on to detail other stories of harassment and assault and takes the viewer up to the 1980s and '90s, including interviews with women who experienced assaults as part of their entrée into the world of entertainment. The "casting couch" culture was alive and well during this era, and the women interviewed detail situations where they were attacked by people in power in the entertainment industry (Zimmerman 2019a).

For example, in one scene, *Rocking the Couch* details how one of the women, Alana Crow, was assaulted by a stage manager on the set of a soap opera but was told by her union representative to do nothing about it, or else she might face being blacklisted by the network (Zimmerman 2019a). Later in his career, the stage manager who assaulted her, Jerry Blumenthal of the soap opera *General Hospital* (ABC, 1963–present), was ultimately fired for sexual harassment after twelve additional women came forward with similar stories of harassment. In another incident recounted in the documentary, Crow was also assaulted by another man, Wallace Kaye, an agent who offered her voiceover acting work. This time, however, she was able to extricate herself when another actress interrupted him by knocking on the door. Crow called her union again, the Screen Actors Guild, and was told to write a letter of complaint this time. Despite writing the letter, the Screen Actors Guild didn't do anything, even though there were several other complaints against Kaye by other women. Kaye was finally arrested and charged with multiple counts of sexual battery when a non-guild actress reported his actions to the Burbank Police Department, after SAG turned her away. The police department set up a sting operation with a female detective posing as an aspiring actress, and the same attempted assault took place.

Both the story of Wallace Kaye and Jerry Blumenthal reveal that incidents of sexual harassment and assault were occurring back in the 1990s and could have led to an earlier version of a #MeToo movement. The industry union, the Screen Actors Guild, or SAG-AFTRA, is indicted in the film for not taking these claims seriously and for trying to minimize them during this period, and were thus guilty of indirectly allowing these actions to continue for several more decades. In other interviews in the documentary there was a sense that being a young actress during that period meant you were subject to having managers try to set you up romantically with male producers in order to get work in the industry (Zimmerman 2019a). The larger point of the documentary is that if other industry players, especially the union that was the main form of labor protection for these young women, had acted on their complaints and took them seriously, many of the current allegations could have been averted.

Docuseries in Real Time: Untouchable

One documentary that covered the fall-out of the Weinstein case directly was called *Untouchable* (Hulu, 2019), created by BAFTA-nominated filmmaker Ursula Macfarlane, who filmed Weinstein's accusers and provided them with the chance to publicly reflect on how they were essentially told to keep their "mouths shut." Some of the women interviewed for this project included Rosanna Arquette, Hope D'Amore and Paz de la Huerta. The narrative of the documentary was told from the perspective of the women who had suffered at the hands of Weinstein. The stories they told shared a common theme of being a young person trying to break into the entertainment industry. They meet Weinstein, a famous moviemaker, and the initial encounter turns into a scene which includes a variety of actions, ranging from threats to intimidation, exhibitionism, sexual assault and violence. The women describe variously feeling trapped, frozen, or dissociating from what was going as they were told by Weinstein that their careers would be ruined if they didn't do as they were told (Verongos 2019). In one scene, Macfarlane plays a recorded voice of Weinstein as he pleads with one of his victims about one of his encounters that eventually ends in a financial settlement and a nondisclosure agreement.

In addition to this documentary, there was another mini-series based on the #MeToo movement called "#MeToo, Now What?," put out by PBS in February 2018. The five-part miniseries explored the effects of the movement and also offered a focus on the ways in which race, class and gender are affected by the culture of harassment. The series, hosted by Zainab Salbi, was an attempt by PBS to look at the factors that have allowed sexual harassment to continue in the workplace, in spite of the fact that there were laws put into place in the 1970s banning this kind of behavior. In their unfolding of the story of sexual harassment, they also bring in a variety of other issues, including gender discrimination, pay inequity, and popular culture, as well as how men could be engaged to be part of the larger discussion. The show ended up featuring individuals from all walks of life, and included celebrities and leaders from a variety of industries to flesh out how sexual harassment has been part of their fields and how to create "lasting cultural change" (PBS).

Documentaries of Sexual Predators

Another set of documentaries that emerged in the wake of #MeToo explored the ways that, like *Rocking the Couch,* sexual assaults that occurred in earlier decades were treated then and now. One documentary, called *Surviving R. Kelly,* was a six-part documentary that aired on the Lifetime channel and which covered the story of the rhythm & blues star R. Kelly and his obsession with and victimization of primarily African American young women. Though allegations of abuse had been lodged against him for the previous three decades, there had been no public denunciation of him, despite the fact that the allegations included young women being kept prisoner in his home and being emotionally and physically abused, as well as being starved and isolated. In 2008, he was brought to trial on fourteen counts of child pornography but was acquitted. Throughout this time, the singer denied all of the charges and many of his fans as well as the team that worked for him continued to remain loyal to him. The first episode of the documentary provides the context for thinking about how popular music has long focused on idealizing relationships with younger women. From songs like Chuck Berry's "Sweet Little Sixteen" to Elvis Presley's and Jerry Lee Lewis's courtships and marriages to significantly younger women, the documentary offers a framework for thinking about how girls have been preyed upon by older musicians.

By offering this context, the viewers are given a perspective that helps to re-frame these relationships in the era of #MeToo. This is especially important because the documentary series, while focusing on the life of R. Kelly and his mistreatment of young girls, is also trying to help the viewer and the audience recognize the way fans of these performers were willing to overlook these problems. By offering the benefit of the doubt, Kelly's fans are challenged for their "cognitive dissonance," as *The New Yorker* writer Briana Younger (2019) describes it, which allowed them to enjoy his music while ignoring the stories of the black girls and women who had come forward through the years with accusations of assault.[9] It's this kind of commentary, as much as the revelations and descriptions of abuse and assault in the documentary itself, that has emerged in the wake of #MeToo and that the various genres of films and television series have allowed both critics and viewers alike to reflect upon.

The documentary also allowed the producer of it, Dream Hampton, to try to reclaim her own emergence as a writer who herself felt culpable for ignoring the abuses she had heard about, which were considered an "open secret" (*The New Yorker* 2019). In addition, because of the documentary, a Georgia DA re-opened an investigation into the allegations of abuse and assault. R. Kelly was subsequently taken into custody on June 6, 2019, in Chicago on 10 charges of aggravated criminal sexual assault connected to allegations that he had sexually abused four women.

Hampton was a music journalist who had done an earlier piece on Kelly, in *Vibe* magazine in 2000, post Kelly's marriage to Aaliyah, an aspiring R & B artist who was only 15 when R. Kelly married her, while he was in his late 20s. In the documentary, young women describe physical and sexual abuse and being kept from their families. In Episode 5, the docuseries shows the rescue of one young woman, Dominique, by her mother Michele Kramer. They showed Kramer looking for

Dominique and though Kramer had called the police, the daughter was over 18 and in a relationship so could not be rescued. In fact, R. Kelly's associates had called the police on the mother by for trying to rescue the daughter. The implication in the documentary was that because all of the young women were also black, the police were less willing to act on the allegations. In this way, the stories of #MeToo are complicated by issues of race and class as contributing factors for why these women were not believed, as well as the fact that the singer R. Kelly was at first believed over the women themselves because he was a celebrity.

In an interview with Gayle King in 2019, R. Kelly asked the question of why the allegations against him were coming out now. To summarize, though many men have made the same complaint that they were being accused years after the alleged abuse took place, this question raised the larger question of why it is that we are now seeing these kinds of accusations come forward in multiple narratives, including docudramas and documentaries. Gayle King offered an answer to Kelly's complaint by responding that the accusations are coming forward "Because we're in a different time when women are speaking out, and now women feel safe saying the things about you" (Kornhaber 2019). It is in this response that we can truly begin to measure the impact of the #MeToo movement in film and television, by noting how this "different time" means that despite the denials, and in Kelly's case, the histrionics he demonstrated in this interview, women's stories are finally coming forward in documentaries, docudramas and other narratives.

Reviewing the Past in Light of the Present: From Nostalgia to a Deeper Truth

In 2019, two more docuseries came out that, while covering events that occurred in the 1990s, reflected the new sensibility about sexual abuse and assault that arose in the #MeToo era. In recent television shows and docudramas, nostalgia for earlier eras, including the 1990s, can be seen in a variety of genres, from the comedies *How I Met Your Mother* (CBS, 2005–2014) to *The Wonder Years* (ABC, 1988–1993) and *GLOW* (Netflix, 2017–present) as well as a number of remakes. In some recent docudramas, however, there is a new focus on some of the victims in this era and a critique of how these victims were originally viewed in the 1990s when the abuse occurred. For example, as can be seen in Lifetime's *Surviving R. Kelly*, the abuse that occurred and that was chronicled by Dream Hampton brought a new level of awareness of the large number of allegations and allowed the victims themselves to tell their stories of what occurred and how the justice system had failed them. In two additional shows, including HBO's *Leaving Neverland* (which premiered January 25 at the 2019 Sundance Film Festival) and *Lorena* (Amazon Prime Video, original air date, February 15, 2019), this theme of revising viewer's original perceptions of key events from the 1990s through hearing victims tell their stories also emerges as a way to understand the differences between how popular culture and society more generally failed these victims during that era.

Amazon Prime's *Lorena*, for which Jordan Peele served as an executive producer, takes a fresh look at a case from the 1990s, when Lorena Bobbitt cut off her husband's penis and threw it out of their car window. The director, Joshua Rofe, revisits this case and trial that occurred 25 years prior, and puts Lorena's act into a specific historical context. He shows that her husband, John Bobbitt, was accused of sexually assaulting her, and he also had been convicted in 1994 of battery against another woman. At the time of the case, however, Lorena's act of cutting off her husband's penis became the object of humor by late night comedians, and by making jokes about it, they ended up minimizing the husband's assaults against her, which had been ongoing throughout their marriage. Though the husband was acquitted of rape charges against Lorena, and she herself was found not guilty because the jury found that it was the result of an "irresistible impulse," the documentary offers a critique of a culture that turned stories of sexual assault into a joke.

The series itself offers the viewer an historical perspective, one which shows how other trials and events at the time that profiled incidents against women, including Nicole Brown Simpson, Anita Hill and Monica Lewinsky, also ignored the sexual misconduct and assault of the men involved. Looking retrospectively at Lorena Bobbitt's court testimonies, the documentary highlights the ways in which the #MeToo movement has influenced our understanding of the case and implicates the audience as well as the media at the time for making a joke out of her trauma. The #MeToo perspective the filmmakers adopt reframes the event and makes it a documentary about the effects of domestic abuse. As a recent immigrant, Lorena was only 24 years old when she cut off her husband's penis, and after her husband psychologically abused and raped her, he forced her to get an abortion. By offering this work of "corrective compassion," to use Megan Garber's (2019) term, the audience is able to understand the experience, through her unedited re-airings of her televised trial, of what it was like to live with this kind of abuse. Noting the power of this testimony and of the audience hearing the voice of Lorena for the first time, Garber offered, "The whole thing is deeply unfunny. And it is testimony that, more than 20 years later, serves as a rebuke not only to John Wayne [Bobbitt], but also to a media system whose first instinct, when learning of Lorena's story, was to laugh" (Garber 2019).

In HBO's *Neverland*, created by Dan Reed, the story of Michael Jackson and his alleged abuse of young boys is told from the perspective of two of the young men, Wade Robson and James Safechuck, who lived with Jackson. Though both men originally denied that any abuse had taken place and stood behind Jackson, the film shows why they didn't feel they could tell the truth when they were first questioned. The film also shows what kind of toll not telling the truth had taken on their own psyches and the lasting damage that the abuse and the silence has had on their lives (Feinberg 2019). The visual impact of watching the docu-series is powerful, with the young men talking directly to the camera and telling the audience that, like many other #MeToo victims, they were deeply ambivalent about coming forward, both because they didn't think they would be believed as well as because they had complicated feelings about their abuser. Michael Jackson was a larger-than-life figure and had tremendous influence over not only their own lives, but the lives of their families, and this kind of power left the men both vulnerable to and in awe of Jackson. They also describe their earlier feelings of love for him, offering a window into the complicated ways that abuse can play out in the psyches of victims even as they come forward to tell their stories. The abuse they describe, however, is both graphic and wrenching, occurring when both are pre-pubescent. To a contemporary audience who has lived through #MeToo, a dominant theme emerges of young adults who were robbed of their childhoods, which poses a stark contrast to our earlier understanding of Michael Jackson as a beloved and revered figure in the 1990s. Describing the impact of these stories for our historical time period, Daniel Feinberg noted that, like *Surviving R. Kelly* and *Lorena*, "all three docuseries are the fruit of a new, darker but also distinctly progressive kind of nostalgia—a hunger to retell and rehear these stories at a moment when we're trying, in theory, to do a better job of believing both men and women when they speak their truth. This is television that shows us how far we've come and perhaps how far we still need to go" (Feinberg 2019).

Another docudrama that dealt with the issue of sexual assault streamed on Netflix in the fall of 2019 is *Unbelievable* (original air date, September 13, 2019). The series was inspired by real events which had been written about in a Pulitzer Prize–winning *Pro Publica* article called "An Unbelievable Story of Rape," as well as a *This American Life* radio episode called "Anatomy of Doubt." The story is about a young woman named Marie Adler (Kaitlyn Dever) from Lynnwood, Washington, who initially reported a violent rape by a masked intruder who broke into her apartment, but she was doubted by the police and was then herself charged with filing a false report. Years later and miles away in Colorado, it would eventually emerge that Marie had been the victim of a serial rapist, and if she had been believed initially, other rapes might have been prevented. The director Lisa Cholodenko, after being sent the *Pro Publica* article, felt that it was already being written in a way that allowed for a compelling docudrama. Describing the timing of the series in relation to the #MeToo movement, the executive producer Sarah Timberman explained that "we started working on this two years ago, and it was obviously relevant and had been, unfortunately, for centuries, but then suddenly the material, and this story in particular, felt like it completely coincided with this wave of consciousness about this subject" (Turchiano 2019).

Although the story focuses on the brutal rapes committed by a serial rapist, and the fact that a young woman was not believed and was instead charged and convicted of filing a false police report when she went back on her story, the aim of the director was not to portray the police as uniformly bad, but to "keep it even-handed, and tell this incredible story as it really was" (Combemale 2019). Part of the way that Cholodenko, in directing *Unbelievable*, tried to convey the experience of the women was to be both objective and not "politically motivated," but to also stay subjective with the experience of trauma they were going through. In addition, she also wanted to show the work the women detectives were doing and how they themselves were experiencing the case and what it meant to finally get an opening in the case. In terms of

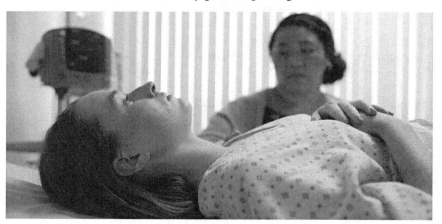

The female body as crime scene: Rape survivor Marie (Kaitlyn Deaver) must cope with repeated assault at the hands of police, doctors and the judicial system in *Unbelievable* (Netflix, 2019).

the story itself, it is clear that it centers on the question of whose story will be believed in the case of a sexual assault. This in turn also raises the question of the difficulties involved in bringing the individual predators who commit sexual assault to justice and the ways in which survivors are afraid to speak up because they will be treated with the same degree of suspicion as their alleged attackers (Blake 2019).

Unbelievable is also revolutionary in the way it chooses to show the actual rapes, not from the perspective of the perpetrator but from the experience of the victim. This is a real departure from what is too often seen on television shows that depict rape, where the predator's point of view is often visually the one that is given by the camera angle used. Susannah Grant, who co-created *Unbelievable*, was clear that she wanted to move away from the conventional portrayals of rape, and noted that "I was conscious of how accustomed to the world of 'rape porn' we all have become—the voyeuristic view of quasi-violent sexuality. It's present in a lot of the images we're exposed to culturally.... I really didn't want to watch a rape. I wanted the viewer to understand the experience of that sort of violation and assault" (Blake 2019). In perhaps one of the more powerful scenes in the film, Marie (Kaitlyn Dever) undergoes a forensic exam after the assault. The camera is clinical in its matter-of-fact portrayal of the nurse who explains to Marie what she is going to do. She then goes ahead and begins swabbing and photographing and collecting evidence in an efficient manner. Marie is silent with her hands held tightly over her chest, and the room itself is quiet except for the sounds of the instruments the audience hears as the nurse does her work, including opening the speculum. Megan Garber noted that while this scene could be potentially voyeuristic, the filmmakers were clear that they wanted to do something else with this rape-kit scene, or as she says, "The scene is invasive. It is not, however, voyeuristic. And that makes it all the more bracing: This is not the kind of moment that is typically shown in depictions of sexual violence. This is *Unbelievable*, revealing what happens when a body becomes a crime scene" (Garber 2019).

Source Citation

Docudrama

Another docudrama television series that dealt with sexual harassment and assault was the seven-part miniseries called *The Loudest Voice* (Showtime), which premiered on June 30, 2019. The series covers the professional and personal life of Roger Ailes, the CEO of Fox News from his founding of it in 1995 to 2017. In 2016, Ailes resigned from Fox News after a series of allegations of sexual misconduct were launched against him by the female employees who worked at Fox News. While the series covered some instances of his predatory and degrading behavior toward women, it was only in the third episode, "2008" (original air date July 14, 2019), that they revealed the full extent of his behavior in his treatment of Laurie Luhn (Annabelle Wallis), a former booker of guests for the Fox News Network. Describing her relationship with Roger Ailes as a form of "psychological torture," Luhn claimed that Ailes sexually harassed her for over two decades (Nicolaou 2019). Though the relationship, as it was depicted in the series, was viewed as consensual on one level, at another level it revealed a kind of power dynamic that allowed Ailes to dominate Luhn's life. Luhn met Ailes at a George H.W. Bush presidential campaign headquarters in Washington, D.C., when she was a 28-year-old former flight attendant and he was a married Republican political strategist close to 50. Ailes then went on to hire Luhn to do intelligence work on other Republican strategists. The relationship from that time on involved the exchange of sexual favors for professional contacts and work, and Ailes eventually hired her to become part of the new Fox News network in 1996.

The series goes into great detail about this specific relationship as a way to provide the context for understanding how Ailes used his power and influence to coerce and manipulate the female employees at Fox News. In Luhn's case, this kind of coercion led to her mental deterioration, in which she displayed paranoid episodes, based in part on the fact that Ailes had her and other employees surveilled at Fox, as well as his keeping sexually explicit photographs of her as a form of blackmail. She eventually further deteriorated mentally and tried to commit suicide, and it was only in 2011, several decades after she and Ailes met, that she finally reported the sexual harassment and settled with Fox for $3.5 million. Luhn also signed an NDA agreement which meant that she couldn't discuss the abuse or go to court against the parent company Fox (Nicolaou 2019). For the showrunner, Alex Metcalf, the real hero of the mini-series, which covered the extensive sexual harassment by Ailes, is revealed in Episode 6, "2015" (original air date August 4, 2019), where the former Fox News host Gretchen Carlson (Naomi Watts) makes the decision to sue Ailes over his sexual harassment of her while she was at Fox News. Carlson was a Fox News co-host of *Fox & Friends* and was subjected to systematic humiliation, belittling and sexual harassment from Ailes while she worked at Fox.

In an early scene in the series, Carlson goes to Ailes to complain about her co-host grabbing her arm while she was on the air to have her stop talking. This kind of behavior was part of a larger pattern at Fox News where gender bias was embedded in the daily life of the women who worked there. Ailes responds to her complaint by telling her that she needed to rise above her initial reaction to this. He admonishes her, "Gretchen, Gretchen, you're Miss America ... how would Miss America handle this? With grace, charm, you'd smile, give a little twirl, wouldn't you? So let's see it. Why don't we see a little Miss America twirl?" (Miller 2019). While on one level, the storyline focused on the ways in which Ailes was responsible for the creation of a kind of news on cable channels that has come to define the political landscape we are living in, closely related to this is the way in which Ailes used his power to intimidate and harass the women who worked there. The sexual culture of the workplace, which eventually led to Ailes being fired for his harassment and assault, was part of the larger story and ultimately led to his downfall.

Another fictional account of sexual harassment by Roger Ailes was the dark comedy and drama *Bombshell* (Lionsgate) which came out in October 2019. The producer of the film, Charlize Theron, hired Nicole Kidman to portray Gretchen Carlson while Theron herself played Megyn Kelly, who had also made sexual harassment allegations against Ailes. While covering much of the same ground as the docuseries *The Loudest Voice*, the aesthetics are somewhat different in that it is shot in a docu-style that is observational (Walsh 2019). In directing the film, Jay Roach described doing extensive research on the events leading up to Aile's resignation, including interviews with several of the women from Fox News who broke their nondisclosure agreements with Fox to provide the filmmakers with background information.[10] Theron herself described her desire to make sure that they didn't turn Kelly into a unidimensional sympathetic character or an inherently good person. Rather, she felt it was important to "always stay on that path to truth. That's what she deserves, that's what the movie deserves, and that was what was most important" (Walsh 2019).

The film has a composite character, Kayla Pospisil (Margot Robbie), an ambitious young woman who wants to move up the ladder to become a news anchor. Popisil is allowed to meet with Ailes (John Lithgow), ostensibly to discuss her potential promotion from her current desk job to become an on-air talent in the Fox News universe. After she enters his office, which she is buzzed into by an older woman who is Ailes' assistant (Holland Taylor), she begins to describe her qualifications to Ailes. He abruptly interrupts her and orders her to "stand up and give me a twirl." Although she is clearly uncomfortable, she tries to normalize it with a nervous smile and does the twirl, at which point he tells her to hike up her skirt, even though she is wearing a mini-dress, with the justification that television is a visual medium. As she raises her dress, the audience hears his breath grow louder and more labored, and he then commands her to keep lifting up her dress until she finally lifts it high enough that he can see her underwear. She ends up in tears when she leaves his office. In explaining what he was trying to get at with this scene, the director Jay Roach explained, "It was about taking something from her, taking her dignity, and then having her feel ashamed enough that she doesn't want anybody to ever know about this" (Lee 2019).

This theft of dignity is part of the way Ailes operated, by being both a cheerleader for the women and, at the same time, a manipulator, forcing them into doing things they didn't want to do, with the threat that he could influence their career in either direction depending on whether they did as he demanded. As Roach pointed out, "That's the weird trap he's laid for her. And the women we interviewed while researching described that feeling. I just wanted to make sure the emotional levels were there and that the empathy you would feel for this is

Bombshell (Lionsgate, 2019): Gretchen Carlson's allegation against Roger Ailes predated MeToo; when it finally came, Ailes' downfall was the subject of multiple docudramas. Shown left to right: Charlize Theron as Megyn Kelly, Nicole Kidman as Gretchen Carlson, and Margot Robbie as Kayla Pospisil.

just devastating. It certainly was for me in the room. I've never filmed anything as excruciating" (Lee 2019). In trying to make the scene as realistic as possible, Charles Randolph, the screenwriter for *Bombshell*, at the same time also thought that he wanted to make sure the scene was dark enough that the audience, including men, could understand what Kayla was going through but not so much that it would be exploitative. When he and the director wanted to "tone it down," at one point, the producers, including actress Charlize Theron, told them not to change a single frame so that the audience could understand the full horror of what the woman was going through. As they did the research for the script, they learned that Ailes himself was not simply a villain, but that he was "Machiavellian," and that he was able to manipulate women by offering paternal advice and making them feel comfortable by being humorous and a good coach. As Roach described, "He would have good advice, and women would come into his office and feel very comfortable at first. Then it would drift to this darker place" (Lee 2019). In classic form, then, the horror is heightened as the perpetrator is at first shown to be friendly and kind and then shifts abruptly into that "darker" place.

Conclusion: Making Sense of Imperfect Heroines

In many ways, as different as comedies and dramas are from one another, there are some common themes that emerge in the #MeToo era in these different genres. In one striking example, the idea that women may to some extent be complicit in the cover up of sexual harassment can be seen in a variety of these shows, including *Bombshell*, as well as the comedy *The Morning Show*, and the dramedy *Succession.* In each of these shows, there is a woman who is ambivalent about coming forward or in risking her own career for the sake of telling the truth about what happened when someone was sexually harassed at work. While it is often portrayed in recent film and television that the abusers are wrong and culpable, the question of whether the women around them may have enabled them or remained silent is more ambiguous territory that is also explored.

In the film *Bombshell*, for example, the character Megyn Kelly finds that, though Gretchen Carlson sues Ailes for sexual harassment, it was only when Kelly came forward that the accusations against Ailes took hold. In the film, Carlson's lawsuit was not taken seriously, and Kelly did not at first come forward to tell her own story of being harassed by Ailes a decade earlier. The reason is that she was in the middle of renegotiating her contract with Fox, and the movie makes it seem as if she were weighing the impact coming forward herself might have on her own career. Like Kelly, the character in HBO's *Succession*, the daughter Shiv Roy (Sarah Snook) also finds herself weighing her own career in Season 2, episode 9, "DC" (original air date October 6, 2019), over allowing a woman in her father's media company to come forward and testify to sexual harassment that occurred in his company. While she is portrayed as the most ethical of the four children, with each of her brothers being morally compromised in various ways, in the second season her father has her confront Kira (Sally Murphy), one of the victims of harassment, directly to ask her not to testify to Congress about what happened. She admits that in order to keep her status in the family and the company, she is willing to compromise her own principles to get the woman to not testify. She does this in part by cleverly manipulating the woman and telling her, "You're going to have a lot of people on your side who will sing your name and back you up. But the other people? The other side? Normal people? ... They'll call you a slut and a whore and a money-grubber.... From tomorrow that's all you'll ever be, to your grandkids, to people you meet on vacation. When they Google you, pages and pages of filth and lies. First line of your obituary, last line." Instead of coming forward, Shiv urges her to work with her from the inside to help her destroy the people who committed the sexual assault against other women in the company. Eliana Dockterman (2019) notes that what Shiv says mirrors much of what happens to women when they come forward to testify, as in the case of Christine Blasey Ford's testimony to Congress against Brett Kavanaugh, which occurred exactly one year earlier to the day that the episode aired. And, as much viewers want to believe that Shiv will indeed hold these abusers to account, it is by no means clear that she will do so.

These tensions about the complicity of women is echoed in several comedy and dramatic series that emerged during this period, in particular in Apple TV+'s *The Morning Show* (original air date November 1, 2019) starring Jennifer Aniston as morning co-anchor Alex who tries to remain loyal to her former co-host, Mitch (Steve Carrell). When Mitch confronts her with the fact that she was aware of his behavior when they worked together, she feigns ignorance. He then accuses her of behaving badly toward these women as well, by mocking them or rolling her eyes at them.[11] At the same time, Mitch himself wants to bring everyone on the show down with him. When one of the women who was a survivor of Mitch's assault ends up taking her own life, the show moves from comedy to a serious moment, where Alex announces on camera that her own head of the network was aware of Mitch's behavior and allowed it to happen because of the toxic work culture fostered there. Rather than show her as completely honest, however, there is the question of whether she is trying to make it look like she is coming forward as an honest broker, or merely protecting her own image. Alex is speaking truth to power, in other words, even as she admits to having remained silent for so long, as when she goes on air to say, "I'm as culpable as anyone in terms of not calling out or not helping to end the sexual misconduct that goes on in this f-cking building" (Dockterman 2019).

As we will see in the next chapter on horror films in the #MeToo era, the ways in which both male and female characters are positioned in these fictional narratives seem to mirror what is happening in the larger culture around trying to make sense of who is culpable and how we move forward once sexual harassment is discovered in the workplace and beyond. Just as in real life, these characters both embody the contradictions of how difficult it is to speak up and at the same time, how critical it is not to do so. And for those who have witnessed it, or knew of its existence, similar questions remain, even if it means challenging their own earlier complicity or jeopardizing their own careers. In these ways, both comedy and drama have tried to bring the discussion forward and allow their viewers to raise these questions as it becomes part of our cultural legacy in the #MeToo era.

6

Anxiety, Terror
and the Mundane

Horror Films and Reality Television

One of the most enduring film genres that explores women's place in society is horror. Speaking more generally about the purpose of horror films for a viewing audience, film theorist Noel Carroll offered that "[t]he genre [of horror] is a means through which the anxieties of an era can be expressed. In horror, characters face monsters that cannot be explained. They are abnormal, outside of our world of reason and meaning" (Thom 2018). For many creators in Hollywood and abroad, the #MeToo movement, with its effort to expose the "monsters" sexually harassing and assaulting individuals, offers ripe material on which the genre of horror can capitalize. Speaking of the ways that #MeToo themes are now showing up in horror film plotlines with increasing frequency, Maren Thom (2018) writes about the latest film in the *Halloween* franchise as a prime example. As she noted, *"Halloween* is a #MeToo movie in other, less obvious ways, too. Horror movies describe not just personal, subjective trauma, but also the symptoms of the ideas that shape society.... Horror monsters are ideological symptoms which, through expressing societal anxieties, show us that these anxieties are ideological in and of themselves."

In this way, contemporary horror films are an important avenue to explore how popular culture is expressing larger cultural preoccupations with issues raised in the wake of the #MeToo movement. Ironically, this focus on the terror of the victims of sexual harassment and assault marks a departure from earlier critiques of horror films as exploiting women. Jessa Crispin (2020), for example, observed that while horror films have retained their popularity through different periods in film history, in recent times they have come under scrutiny for the ways in which they feature women as victims of exploitation, torture and murder. In many ways, the #MeToo movement has made these questions even more relevant, and filmmakers have addressed this by incorporating #MeToo themes into their storylines from the woman's point of view. There have been some inroads in terms of how these stories are told, especially when there are women filmmakers and directors and writers and producers creating these stories.

Crispin offered a helpful breakdown of the types of female characters that have emerged in the wake of the #MeToo movement. She described one protagonist as the "pure victim," and a second as the "strong heroine." The third version is the pure victim who becomes the strong heroine as a result of having to endure some kind of violent experience (Crispin 2020). While the films themselves may not necessarily be more nuanced than more conventional horror films with "damsels in distress," they are nevertheless revealing for the ways in which they have changed the conversation around what kinds of stories are told and whether the protagonists' perspectives are foregrounded in the story. Films such as *The Assistant* (release date January 21, 2020), an indie film that came out three months after Weinstein was convicted, *The Invisible Man* (2020), and *Black Christmas* (2020) are just some of the examples of these new horror films. *Promising Young Woman* (2020), another indie film like *The Assistant* (2020), is yet another horror film/psychological thriller that has emerged in the wake of the #MeToo movement, and has the main protagonist, played by Carey Mulligan, avenging women's victimization.

What follows, then, are two examples of contemporary films that have drawn on the genre of horror and that have #MeToo themes. In *The Assistant* (2020), there is the example of a young woman who observes the victimization of other women and tries to put a stop to it. In yet another film, *Ma* (2019), we see an example of a woman who was herself victimized who goes on to victimize others as a result of her PTSD from being sexually coerced. The complicated sexual dynamics, as will be shown, that come about as a result of sexual harassment or exploitation often have unintended and multiplying effects in the horror genre that filmmakers are eager to capitalize on in order to have their material feel current and express contemporary anxieties over issues of sexual assault.

The Assistant *(2020)*

The Assistant (2020), written and directed by Australian Kitty Green, has a relatively simple storyline which shows the routine tasks repeated by the main character, a recently hired assistant to a large film production company. The young woman, Jane (Julia Garner), is an assistant to a film mogul whose behavior holds a striking resemblance to Harvey Weinstein. Kitty Green spent a year interviewing people from Miramax (where Harvey Weinstein was the boss) about the work culture there. She interviewed several administrative assistants and was able to demonstrate at the granular level how power and intimidation work and how it was that so many people accepted this behavior for so long. These include such micro moments such as who gets in and out of the elevator first, which becomes a central portal in the film to signal who has more power and who is clearly lower level. The actress Julia Garner is central to making the film hold the kind of terror that eventually accumulates for the viewer, as we watch Jane's face go from being relatively calm and thoughtful to an increasing sense of anxiety and finally frustration and then resignation, all without having the character actually explain what the issues are to viewers.

While the young woman, a recent college graduate, has only been at the job for five weeks, it's clear that she has already absorbed the office politics and dynamics that she is trying to scrupulously follow. The boss is rarely seen on screen; in a few instances he is shown in a meeting, but the camera is too far away to see him clearly. And, while you can't hear the words he is saying, you can hear the tone of his voice, when he is in a meeting or when he is entertaining someone or when he is angry. In this way, he is ubiquitous in the film as a kind of presence that could appear (albeit in another room, from where the camera is focused) at any moment. From the opening scenes, then, *The Assistant*, while seemingly a movie about workplace culture, also conveys the atmosphere of a horror film, where the threat is presented in the form of a boss who wields absolute power even when he can't be seen or his words be understood. In fact, the audience never learns his name because he is referred to throughout the film as "he," as if his name is not necessary to understand the kind of power he wields over the whole office.

As the newest assistant to him, in a short amount of time, Jane is already accustomed to doing all of the grunt work that is expected in this environment. No task is too small or outside of her responsibility, from wiping up the crumbs on the table and removing food after a meeting, to making travel arrangements for the boss and his entourage, to taking calls from the various women in his life who are demanding to speak to him. She is shown in the first scenes of the film leaving very early from her apartment in one of the boroughs of Manhattan and taking an Uber to the office building in lower Manhattan for what will be at least a twelve-hour workday. As the first to arrive in the office while it

The Assistant: **The 2019 indie drama starring Julia Gardner echoed the Weinstein case and detailed how power and intimidation work to silence those surrounding an abuser (Bleeker Street).**

is still dark out, she sets about doing all of the housekeeping and paperwork. These quotidian tasks include sanitizing the couch in the boss' office where she finds an earring on the floor right next to it, presumably from a woman who had sex with the boss on the couch the day before. She picks it up and continues with her tasks, Xeroxing scripts that need to be read for upcoming projects, washing the dishes in the office kitchen and ordering and delivering food to the boss and other assistants in the office. The whole film takes place during a very long day, and Jane becomes increasingly concerned with what she is beginning to understand about the sexual dynamics at play in the office. While this entry level job is seen as a great opportunity for her to get a break in the film business and become a film producer, she quickly learns that it comes with a heavy moral dilemma, which is to basically ignore the sexual predations of her boss. At one point in the film she will be told not to worry about his behavior because she is not "his type," as it was assumed that the only reason she would bring it to anyone else's attention was because she was afraid for her own safety.

The whole film is shot from the point of view of Jane, which both foregrounds her experience at the same time it creates an increasing sense of dread in the viewer. At one level it is a film about a corrupt workplace, where everyone is asked to cover for the boss and excuse his transgressions and angry outbursts. At another level, however, it is about the increasing levels of degradation to which Jane is subjected. This is revealed when the boss screams at her over the phone on two different occasions and she responds by quickly composing two different emails apologizing for her supposed transgressions. As the day progresses, the details about her boss and his actions around women in particular become more threatening and at the same time, remain unclear, since all of the action that he engages in happens off-screen and away from the audience's view. At one point, the boss is taking a meeting with a very young woman he met while she was working as a waitress at a ski resort. The naïve young woman (Kristine Foseth) is from Boise, Idaho, and the boss flew her to New York where Jane had to pick her up at the airport and take her to "The Mark," a fancy hotel in the city. Later, we learn that the boss goes to the hotel for the afternoon and then the next day, the young woman is shown at the office, where she explains that she is the new assistant to him. Jane is supposed to "break her in," but it is clear that there is little for her to do, and Jane tells her that she can go home early.

In another scene, Jane sees a young woman, presumably an actress, waiting to take a meeting with the boss. Like the other young woman, she is beautiful and in this case, it is clear that she is hoping to get a role in an upcoming project the boss is doing. In a different scene, we see Jane waiting at the elevator and when the door opens, there is yet another beautiful woman to whom she gives the missing earring, indicating that she was the one who had sex with the boss the previous day. The couch becomes a central focus of the film, as it symbolizes the boss' sexual exploits with different young women. This is reinforced in another scene, where several male executives enter the boss' office to take a meeting and they make a joke about not wanting to sit on that couch, presumably because they know the boss has sex on it. Everyone, in other words, is seemingly in on the "joke" that the boss is a sexual predator and finds it funny rather than inappropriate or troubling. At a later point in the film, another female assistant tells Jane that a young woman who is taking a meeting with the boss will get much more from the meeting than the boss will, implying that the young woman's career will get a big boost from having sex with the boss. Clearly, all of the other employees have accepted the boss's behavior as part of the cost of working there.

These interludes with young women occur throughout the film and generate an increasing sense of alarm for Jane. In another scene Jane is speaking to a woman who is in L.A. and who is supposed to meet the boss, who is flying out there later that day. Jane tells her that she will be taking the meeting with the boss at his hotel, and the woman doesn't seem to react with any surprise at the location. While the young women around her seem accepting of her boss and his demands, Jane finds herself increasingly threatened by his abusive behavior toward her. At one point, he screams at her over the phone for taking a call from his wife and telling the wife that he would call her back. This enrages him and he verbally abuses Jane on the phone which causes her to hurriedly write an email apologizing to him. At another point, when a similar outburst occurs by him on the phone and she again apologizes, he sends a follow up email telling her how wonderful she is. And, in another scene when she is talking to a driver for the boss, the driver tells her that the boss thinks that she is really smart and you can see the sense of relief in her eyes. In these ways, the mundane kind of terror he inspires as an almost cartoon version of a boss from hell gets upended by these moments where she receives his praise. The overall effect of this is to see him as an abuser who knows how to manipulate the people around him.

This is evidenced most clearly in a scene with Wilcock (Matthew Macfadyen) who works in human resources. Jane finally summons up the courage to go to HR to report her boss. She is interviewed by Wilcock about what she saw, culminating in her description of picking up a very young woman from the airport and taking her to a hotel where her boss meets with the woman and then her showing up the next day as the new assistant. In a tense exchange, the HR representative dryly tells her that she hadn't really seen anything and was inventing possible scenarios of which she had no proof. He then continued to question her along these lines about what she actually did and didn't see and ended the meeting by telling her that she has this great opportunity for her career to be working where she is and questions if she really wants him to file a complaint against her boss. She responds by saying no, and leaves the office, clearly shaken and realizing that the HR representative also knew what the boss had been doing with the young women the whole time and had never taken any action against him.

To summarize, one of the ways that *The Assistant* brings home the problems for women in the #MeToo era is to show how everyone in the office had become complicit in enabling the boss' behavior. Like the other employees around her, Jane quickly learns to focus on how to protect her boss from any intrusions and to cover up any potential ways he may be challenged. The film also focuses on the ways that Jane, as a woman, is tasked with very specific forms of labor, including the domestic labor of cleaning up the conference rooms after meetings, or taking care of the boss's children when another assistant brings them to the office. She is literally his personal assistant and takes personal calls for him from the wife, who angrily challenges her when she finds out that her credit cards have been canceled. She also, and this is where the grooming and enabling behavior comes to the fore, is asked to escort the young woman to the hotel, even as she realizes that the woman was flown in from Idaho to not simply be her boss's assistant but to sleep with him.

At the same time, Jane herself is dehumanized as no one in the film actually uses her name or speaks to her. She is basically ignored and treated like she doesn't exist except when someone needs something from her. When one of the other assistants wants to get her attention, he actually throws a piece of balled-up paper at her and, in another scene, two female employees are talking in the kitchen where she is cleaning up and don't even acknowledge her presence, except to put two coffee mugs into the sink for her to wash up. In this climate, when she finally does get a word of praise or an acknowledgment of her work, she is primed to give her loyalty to the boss since she has been so starved for recognition. As Sophie Gilbert (2020) points out, Green was able, through this character study of Jane, to show how the system works to enable and protect Jane's boss and how Jane has been socialized to accept the behaviors of her boss even as she registers her discomfort with them. Green is able to show that the process of desensitization that Jane undergoes from her mistreatment in turn allows the kind of sexual harassment of other women to be tolerated at some level by Jane and the other employees. As Gilbert (2020) noted, "Already she's [Jane] habituated to the remnants of abuse—the stains on fabric, the bloodied syringes on her desk, the jewelry inadvertently discarded. Getting used to the degradation of the women cycling through the office, and disposing of them too, is just another kind of cleanup." When Jane finally does get up the nerve to voice her concerns, the HR manager challenges her and comes up with a series of reasons why she is coming forward, in effect gaslighting her, and throwing the notes he has just taken into the trash, which suggests his own complicity as well.

In being interviewed for the film, Green explained that she hadn't set out to tell a story about Harvey Weinstein, but rather to look at the ways in which women are treated in the film industry as well as other industries. Her intent was to show that it wasn't so much an individual case of a bad actor, but that the ways in which women are mistreated are more systematic and pervasive. As she set out to interview women she noted that "[w]hat was shocking was how common and prevalent these stories were. People in lots of different industries were feeling crushed by the system. In the entertainment industry, we're getting rid of all these bad apples, firing all these men who are predators, but there are more systemic problems. There aren't enough women in the film industry, and I think the film industry is inherently structured against women" (Schama 2020). And, even though the film wasn't made with Weinstein as the focus, Green was clear that it was only in the wake of the #MeToo movement that she could even get the film made. In addition to difficulties getting financing for it, Green pointed out that people didn't necessarily know what was happening in "those back rooms," whereas now it was clear to the viewing audience what was going on and so, as a filmmaker she didn't have to show it. This knowledge, in turn, allowed her to film the story from Jane's perspective, and the viewing audience was therefore also able to see the system of oppression from her eyes as a woman who has no power. The horror of the film lies not only in her realization of what her boss is doing with these young women, but her own degradation as she becomes, in a sense, an accomplice in his predations against these women. The horror is only lifted when she leaves the job and finally walks away from the toxic work culture she could not change but only flee from, while leaving the other young women behind, presumably to fend for themselves as the cycle of abuse continues unabated.

Ma *(2019)*

While *The Assistant* is not a conventional horror film, in the sense that there is not someone stalking and trying to kill someone, there are other black comedies and horror films that center on revenge and horror that reveal concerns that have been raised in the #MeToo era. For example, another theme that has emerged in the wake of the #MeToo movement and that can be seen in recent films centers on the idea that the predator herself suffered from some form of sexual harassment or assault. Whereas it is usually men who are the predators in horror films, in some recent films it is a woman who becomes a predator because she is trying to take revenge on those she deems as responsible for the trauma. This is not to say there is not a long tradition of films where women try to get revenge on men who have done them wrong, such as *John Tucker Must Die* (2006), which was relatively lighthearted, to *Kill Bill* (Quentin Tarantino, 2003), whose character of the Bride rips out the tongue of the man who had sexually assaulted her. A recent black comedy thriller film called *Promising Young Woman* (2020), starring Carey Mulligan and directed by *Killing Eve*'s showrunner Emerald Fennell, also has this theme of a woman enacting revenge on men due to her earlier history of having been sexually traumatized. Mulligan's character, who dropped out of medical school, goes to a club every week and, when there, "act[s] like I'm too drunk to stand. And every week a nice guy comes over to see if I'm okay." When they then try to sleep with her, knowing she is drunk, she immediately sobers up and asks them what they are doing and exacts various forms of revenge. Emma Specter (2019) noted that, like other female-vigilante films, this film has a lot in common with *Hard Candy* (2005) and *Jennifer's Body* (2009), but there is a distinctly feminist take on it, as part of the post–#MeToo culture. She concluded that "[i]n a year when zero female directors were nominated for Golden Globes, the prospect of a woman telling an unflinching, energetic story about what justice for sexual assault survivors means in 2019 feels just right" (Specter 2019). Fennell herself pointed out that while every woman she knows has experienced some degree of sexual assault, she herself didn't know any men who were sexual assaulters, and part of her quest was to try to understand the "gap of experience and empathy between men and women" (Specter 2019). She also wanted the film to be more realistic than conventional revenge movies, and she wanted to think about what a "real woman might do," rather than a *Kill Bill* scenario which involves a massive amount of violence.

Like *Promising Young Woman*, the 2019 horror film *Ma*, starring Octavia Spencer as the title character, also has a woman who is a predator, and her desire to enact revenge is a consequence of having been sexually mistreated when she was in high school. Spencer, who won an Oscar for her work in *The Help* (2011), was given the title role in the film, which was written by Scotty Landes and directed by Tate Taylor, who also directed *The Help*. In *Ma*, the sexual abuse the title character has undergone is only revealed at the end of the film, though there are hints of it throughout in a series of flashbacks. The movie is set in a small town in Ohio, where a young woman, Maggie (Diana Silvers), is a sixteen-year-old new high school student whose mother (Juliette Lewis) turns out to have grown up there but had moved to California, got married, had Maggie, and then got divorced. She moved back to the town and is now working in the local casino as a cocktail waitress. In the early part of the film, Maggie is invited by the popular girl in the school, Haley (McKaley Miller), to go to a party the next night. While the party doesn't happen, she instead finds herself in a van with Haley and three boys, including one named Andy (Corey Fogelmanis). Each of them take turns trying to get an adult to buy them liquor outside a liquor store, and they are rebuffed by everyone except a middle-aged woman named Sue Ann.

Sue Ann, who is African American and working class, realizes that the van the kids are driving is owned by a local contractor, Ben (Luke Evans), who she knew in high school and who she had a secret crush on. She advises the kids to go to a well-known drinking spot outside of town so they won't get into trouble and then looks up Ben online and tells him that his son is drinking. Ben calls the cops, who go there and scare the kids by threatening to arrest them if they don't disperse. Instead of suspecting that she may have contacted the father, the teens ask Sue Ann to buy them liquor again, and this time she invites them to her basement under the pretense that they won't get into trouble like they did when they drank outside and were caught by the police. The basement quickly turns into the place where teens from the high school go to party, and while the film has recurring flashbacks that show Sue Ann as a lonely teenager in high school, it is not yet apparent what the connection is between her invitation to the teens to make her basement a party space and her past life.

As the film unfolds, it becomes clear that Sue Ann, now nicknamed "Ma" by the teens, once had a traumatic experience with Ben, and she is intent on exacting revenge by terrorizing several of the teens, whose parents had also gone to her high school. The teens begin to realize something is wrong when she waves a gun at one of the boys and tells him to strip down. After he does so, she laughs and says, "I'm not Medea!" and claims that the gun doesn't even have bullets in it. Another time, when the girls go upstairs to the bathroom, despite "Ma's" rule that they never come upstairs, she jumps out of nowhere to tell them that they violated the rules. Another time she encourages Maggie to drink shots, and Maggie wakes up the next morning not knowing how she got home and missing the earrings her mother had given her. The film audience, however, sees that Ma gave Maggie some kind of tranquilizer she had stolen from the Veterinary clinic where she worked. At that point, Maggie realizes that Ma is mentally disturbed and tries to warn her friends not to go there, especially Andy. The terror escalates as the young people begin to realize that Ma is troubled, as she texts them furiously and sends videos asking where they are. In an attempt to win back their sympathies, she eventually tells them she has pancreatic cancer, which works for Andy, whose own mother had earlier died of cancer. He goes back to the basement with the other kids, and Maggie ends up going after him.

As the film unfolds, it becomes clear why Ma wants to get revenge on Andy's father, Ben. When Ma was in high school, Ben and his other friends, the popular girls, tricked Sue Ann into performing fellatio on a young man who she thought was Ben, whom she had a crush on. However, when she got out of the closet with who she thought was Ben, it turned out that Ben was outside of the closet with the popular girls. This act of sexual coercion and humiliation left a lasting trauma for Sue Ann which she resolves by terrorizing the children, and eventually Ben as well. At a later point in the film, she ends up killing Ben by giving him a dog-blood transfusion. At another point, she acts out on her troubled past by keeping her own daughter locked up and sick in the upstairs part of her house, with the implication being that she was so traumatized in high school that she doesn't want her daughter to be similarly subjected to any potential sexual harm.

To summarize, the horror thriller film *Ma* draws on the themes of #MeToo as a way to capitalize on the subject matter. Film critics like Richard Brody of *The New Yorker* believe that the film has no artistic interest, and without the #MeToo plot, might "slip quickly down the drain of the movie market" (Brody 2019). By using sexual abuse as the "big reveal," however, for why Sue Ann invites the local teenagers to her basement and tortures them, the film draws on the issues raised in #MeToo of women who have been traumatized by earlier experiences of sexual abuse. For Brody, this decision to have Sue Ann be troubled and have repressed rage ultimately explained by her experience with sexual abuse, in fact, bears little resemblance to the women in real life who have come forward in the wake of #MeToo to tell their stories and ultimately sensationalizes their experiences. For Brody, this is the real crime because, "in the end, these horror films' shocks are delivered at the expense of the genuine shocks that the underlying stories should themselves suffice to arouse" (Brody 2019).

Halloween *(2018)*

The theme of trauma and abuse is underscored not only as the big reveal in *Ma*, but in other recent horror films that draw on #MeToo themes. While there are nine sequels and remakes of the original *Halloween* (1978), the latest version (Miramax, 2018) also capitalizes on #MeToo themes to drive its plot. In this way, the director David Gordon Green and co-screenwriters Danny McBride and Jeff Fradley made a film which essentially elides the fact that there have been these earlier remakes and sequels. This allows them to offer a version of John Carpenter's original film that shows the babysitter Laurie Strode (Jamie Lee Curtis) as a woman forty years older who is still suffering from the impact of her experience being pursued and stalked by masked serial killer Michael Myers (Nick Castle; stuntman, James Jude Courtney). When the original Myers had gone through her hometown of Haddonfield, Illinois, forty years earlier, he was a masked killer who rampaged and murdered and traumatized Laurie to such an extent that as a grandmother she is still suffering from PTSD. She has become a survivalist who has outfitted her house with booby-traps to protect in the instance that Myers might break out of the mental hospital, Smith's Grove Asylum, where he has been since his initial rampage. Under the guise of allowing himself to be interviewed by two British journalists about his past acts, he ends up escaping from the hospital after killing the two interviewers and then goes after Laurie on Halloween.

In this trauma-themed version of a #MeToo horror film, Laurie is now the avenging heroine after having been victimized by Myers forty years earlier. However, the trauma was so great that she ended up becoming paranoid and loses custody of her daughter Karen (Judy Greer) because she had trained her daughter to be similarly armed and ready if Myers ever escaped. The intergenerational trauma theme that is enacted in Laurie's life extends to her granddaughter Allyson (Andi Matichak), who is alienated from her grandmother because of her grandmother's obsession with Myers. This is revealed during a graduation dinner for the granddaughter that was supposed to be a reconciliation, but ended up with Laurie having a breakdown. On another level, having Jamie Lee Curtis re-enact the role that she played originally at the age of 19 allows the audience to see the horror of having been pursued as a young woman, and the devastating impact this has had on her life. At the same time, this version of *Halloween* allows the viewer the cathartic experience of watching the three women (including the daughter and granddaughter who eventually come together with Laurie to fight Myers) bond over their common enemy, a male predator. At the end of the film, the three Strode women end up vanquishing Myers in the woods. In this way, Laurie goes from being the victim of Myers earlier predation to being an avenger.

Jamie Lee Curtis offered her perspective on the changes her character underwent from the original *Halloween* version by describing the older Laurie as a "#MeToo-era hero" (*Rolling Stone* 2018). In her reading of the newest version of the film she speculated that "[e]verybody is talking about past trauma: Burying it, hiding it, squishing it, silencing it, shutting it up…. It's amazing that this is the world. We're talking about this movie that actually, at its core, is about trauma, and trying to put a real face on horrific trauma, and that is what we attempted to do" (*Rolling Stone* 2018). Other commentators have also pointed to the ways the film tries to evoke the themes of the #MeToo era, and in so doing, updates the classic slasher horror film by imbuing it with the ultimate "back story" of a woman who is so traumatized by her earlier experiences that it organizes the rest of her life around this central event. Commenting on the ways this version takes on the themes of the #MeToo movement, Peter Travers noted that "it's the troubled times that we live in that allows this energizing, elemental horror film to touch a raw nerve for #MeToo. We watch a woman call a male monster to account for her own lasting trauma. That's too real to laugh off as Hollywood make-believe. We're living it" (Travers 2018).

Unsane *(2018)*

There have been several other films and television series that take up the theme of the trauma created by a sexually charged predator as the explanation and backdrop for the unfolding of the storyline. Another film that can be considered a psychological horror-thriller and which contains themes that have resonated with viewers after the #MeToo movement is Steven Soderbergh's *Unsane* (20th Century–Fox, 2018). The film is innovative in that, while Soderbergh has been making films for over thirty years, he decided to shoot the entire film on an iPhone 7 Plus. This created a kind of experimental visual style, which adds to the mystery of the story of a woman who gets committed to a mental hospital. The film stars Claire Foy as Sawyer Valentini, and in the first scenes we see her working at a new job where she is praised by her boss for her analytical skills. In the next breath, the boss asks her to go to a conference in another city with him for two days, and it is clear that he is basically coercing her to travel with him in order to keep her job. In the next scene, we see her talking to her mother (played by Amy Irving, in a nod to Irving's earlier horror film role in *Carrie*), and she tells her that she really likes her new job and the new city she lives in, which seems disingenuous because she clearly felt threatened by her boss. Sawyer also tells her mother that she left Boston (where her mother is) because she wanted a new start, without explaining why she needed a new start. In the next scene she goes to a bar where she meets a man and she asks him to go home with her, but after they kiss, she screams at him that she doesn't want this and locks herself into her bathroom and takes some kind of medication. These early scenes signal for the viewer that Sawyer has suffered some kind of trauma around men, but it isn't explained until the next scene, where she looks for a support group for women who have been stalked. We then learn that the reason she left Boston and moved hundreds of miles away was because she was being stalked by David Strine (Joshua Leonard), a man she got a restraining order against but who kept pursuing her nonetheless. As it turns out, the place where she goes to talk to a counselor, a psychiatric institution called Highland Creek, ends up committing her to staying at the hospital overnight. When she protests that she is being unlawfully held, they retaliate by forcing her to stay for seven days.

While she is there she talks to a fellow inmate named Nate Hoffman (Jay Pharoah), who tells her that the institution is running an insurance scam and that they make money from the insurance companies by having patients admitted. After seven days, the insurance stops paying, so he tells her that she should just lay low and wait it out. However, she soon realizes that the stalker has gotten a job at the mental institution and has taken on a fake identity as a nurse. For the audience, however, we are not sure whether this is simply her paranoia or whether this is really David, her stalker. As the rest of the film unfolds, we realize that it is indeed the stalker, and he ends up killing Nate as well as Sawyer's mother and tries to get Sawyer to come away with him to a cabin in the woods. After a series of horror-filled moments of Sawyer trying to get away from him, she finally is able to escape by pretending she is dead and ends up killing him. The last scene of the film takes place six months later. While eating lunch with a colleague, out of the corner of her eye, Sawyer sees a man with his back to her who looks exactly like David, and the soundtrack makes it sound like it is his voice. She takes a sharp knife from her table and she walks toward his table and is about to stab him when he turns around and she realizes that it is a stranger, not him. The camera centers on her face, which is confused, and then she drops the knife and the film ends.

Although *Unsane* was directed, shot and edited just before the Harvey Weinstein scandal broke, the themes resonate powerfully with the #MeToo movement. One of these themes centers around whether a woman would be believed. For the first third of the film it is not clear whether Sawyer is telling the truth that her stalker was in the mental institution, and much of the rest of the film is centered on how to make her voice heard. We later learn that the person whose identity David stole, George Shaw, is found as a corpse in the woods, but this is only toward the end of the film. Sawyer is traumatized throughout the film, first as a result of being pursued relentlessly for two years (including having her home broken into and being texted and called constantly by her abuser). The way the film is shot helps to amplify the way in which we as viewers are invited to experience the trauma and sense of isolation she feels, as the point of view shots are made from Sawyer's perspective, but it is still unclear whether she is really experiencing the images she sees or whether they are some kind of a delusion. Soraya Nadia McDonald observed that "[t]he central question of *Unsane* is supposed to be whether Sawyer is actually being stalked or whether she's a victim of her own paranoid delusions. But an America in which Harvey Weinstein gaslights Rose McGowan with ex–Mossad agents has rendered that question moot" (McDonald 2018). This raises the point that after the #MeToo movement, the idea of a woman being paranoid itself has come to be challenged, after so many women were not believed for such a long time. Instead, we may now be watching movies with a different lens. As McDonald concluded: "Is it even possible now to see a woman on screen being gaslit by a man without thinking about other women who have been silenced and discredited by being labeled as hysterical?" (McDonald 2018).

Reality Shows and the Question of #MeToo

Whereas horror is a form of narrative storytelling that is fictional in nature, the genre of reality television, by contrast, is supposedly based on the truth. And, if the genre of horror is revelatory in terms of how we are currently processing our deeper social anxieties about the world and the unknown, then reality television is arguably offering its audiences some version of the truth or reality that they know. At the same time, reality television, like horror films and television series, also offers an escape for their viewers, one which doesn't challenge their deeper assumptions about the world and human relationships. After #MeToo, however, society is beginning to question these very assumptions about relationships between the sexes, among other things. As Kelly Lawler observed about the power of reality television shows about love and romance, for example, "But as much as these shows are positioned as simple escapes into a world of romance and sex, society is (finally) not treating those subjects as 'simple' anymore. Seeing them dumbed down in reality TV is a weak representation" (Lawler 2018a). And, while the assumption is that reality television shows are somehow "real," they are often as scripted as the most fictional horror stories, and in particular, those shows where romance is involved, the reality that is being portrayed is oftentimes carefully rehearsed for the purpose of gaining high ratings. Despite their fictional construction of supposedly "real situations," however, reality dating shows unintentionally have moments where the truth in fact breaks through. Some of these moments of truth also unfortunately include sexual harassment that took place within their carefully choreographed world.

More generally, there is a whole genre of reality shows that center on dating and finding the right partner that can create situations that are sexually ambiguous and can lead to sexual coercion or harassment. ABC's *The Bachelor* franchise (2002–present), for example, is premised on a storyline about a bachelor or bachelorette and the people vying for their affection. These contestants often have some kind of emotional baggage they have to confront in order to have a meaningful relationship, which adds to the dramatic tension of the show. In the 2019 *Bachelor*, however, the stakes were raised considerably when one of the female contestants, Caelynn Miller-Keyes, shared with the Bachelor Colton Underwood that when she was in college she had been drugged and rendered unconscious and was then sexually assaulted. She told Colton, "[that night] is something that will always be a part of me…. It's the most difficult thing in the world. It's so painful and it screws up every ounce of you" (Lemiski 2019). Caelynn's admission of what happened to her made for an episode that is seldom seen on the show, and as she added later, "It's not one episode on *The Bachelor*, it's not a hashtag. It is truly a movement" (Lemiski 2019). When she first raised the incident, the show offered a Public Service Announcement where they displayed a notice of who the viewer can contact if they had had this experience, the Rape, Abuse, and Incest National Network (RAINN). In this way, *The Bachelor* was trying to demonstrate that it was sensitive to the questions raised by the #MeToo movement around consent and assault.

This public service announcement of *The Bachelor* is offset, however, by larger questions raised by reality dating shows. For example, *The Bachelorette* has been on for 15 seasons, and a new spinoff *Bachelor in Paradise* came on in 2018. The shows themselves often shame the female contestants, as in the season finale of *The Bachelor* in 2018 when the bachelor, Arie Luyendyk, Jr., humiliated the two last women as he made his final choice. In *The Bachelorette* season itself, it was also found out that one of the contestants, Lincoln Adim, had been convicted of indecent assault and battery before he had ever appeared on the show. In *The Proposal* (ABC, 2018–2019), another reality dating show where men and women try to compete for an instant proposal, ABC pulled an episode when they found out that another male contestant had been accused of facilitating sexual assault (Lawler 2018a). These incidents and controversies stand out all the more in the #MeToo era and beg the question of the ethics of putting complete strangers together in sexually provocative situations. In terms of the shows themselves, it is by no means clear that their plots of living "happily ever after" are even true, as the majority of the couples who get together have since broken up. In addition, other reality dating shows such as *A Shot at Love with Tila Tequila* (MTV, 2007–2008) as well as *Dating Naked* (VH1, 2014–2016) also put men and women in potentially vulnerable situations including encouraging alcohol use, or ramping up the conflicts that happen, which can lead to exploiting the contestants and laying bare their vulnerabilities in a sexually charged atmosphere (Lawler 2018a).

What impact, if any, has #MeToo had on the dynamics of these and other reality shows? There was one incident where *Bachelor in Paradise* had to be halted because there was a concern by the producer that there had been nonconsensual sex between two contestants, Corinne Olympios and DeMario Jackson. The series then had the contestants discuss the question of consent in an on-air episode, which was itself staged and contrived because the show itself wasn't able to discuss the nuanced ways that sexual harassment and assault can sometimes occur. In another example, the reality show *Survivor* (CBS, 2002–present), which revolves around contestants being voted off of the show, one of the female contestants, Kellee Kim, was voted off after she complained that one of the other competitors, Dan Spilo, had touched her inappropriately (Bradley 2019b). In the premiere of the 2019 season, Kim and another contestant, Molly Byman, had initially lodged a complaint about Spilo that he didn't have a sense of physical boundaries with them. As more reality programs deal with this issue, it becomes more common for contestants to speak up about boundary issues. In 2018, the CBS show *Big Brother* (2000–present) also had to deal with inappropriate behavior, in particular by one contestant named JC Mounduix, who behaved inappropriately, including opening a bathroom door while one of the contestants was using it and kissing another contestant's armpit and touching him while he was sleeping. Though the producers warned Mounduix many times, he was not removed from the show. In other cases, however, producers have been more responsive and have removed cast members and in general, are trying to be clear with contestants about appropriate boundaries. The executive producer of *Survivor*, Jeff Probst, for example, said that "*Survivor* is a microcosm for our real world ... situations just like this one are playing out in offices and bars and colleges across the country and the world" (Bradley 2019b).

But the stakes are higher on reality television shows where there is money and publicity that plays into the mix. That being said, the producers of *Survivor* were well aware of Spilo's conduct, both because other female players had described the same behaviors as well as because the contestants are monitored by the producers 24 hours a day. It was also clear that they knew that Kim was upset with Spilo's behavior, as when she and another contestant described him as "really touchy" and that he "does not know personal space" (Bradley 2019b). Another contestant, Missy Byrd, described a situation where "At the merge feast last night.... I feel someone wiggling my toes, and I'm like, I wonder who it could be? And it's him.... It's inappropriate touching; I'm not an object" (Bradley

2019b). Kim realizes that it's a pattern with Spilo, and at that point, the producers of *Survivor* break into the interview and tell Kim, "You know, if there are issues to the point where things need to happen, come to me and I will make sure that stops. Because I don't want anyone feeling uncomfortable.... I just want to make sure. This is not ... it's not okay" (Bradley 2019b). All of the players were warned about respecting personal boundaries, and Spilo in particular was cautioned not to engage in any more inappropriate behavior. However, it was then revealed that some of the other female contestants ending up apologizing for backing up Kim in her accusations against Spilo, and they then voted Kim off of the island.

What these subsequent events mean is that, among other things, the contestants themselves have limited knowledge about the larger game, and part of the dynamic of the show is to question the other contestant's motivations and stories they tell. The producers, on the other hand, have all of the knowledge. So when Kim came forward with her allegations against Spilo, it wasn't clear to the other contestants whether she was telling the truth or whether she was manipulating the situation to get Spilo voted off of the island. She wasn't believed by some of them, and so she was voted off the island. The fact that the producers of *Survivor* tried to show how concerned they were about inappropriate behavior by contestants ends up mirroring the strategy other reality shows have employed in the wake of allegations of sexual misconduct. For *Bachelor in Paradise* (ABC, 2014–present), after a production break in 2017, the show had the contestants gather to talk about the issue of consent, and thus made themselves look like they were behaving ethically, despite going back to filming relatively quickly after the harassment incident. At the same time, when Mounduix was warned multiple times on CBS's *Big Brother*, they still didn't kick him off of the show, which allowed them to keep a contestant who was able to draw high ratings. Similarly, *Survivor* ended up eliminating the female contestant who came forward with credible accusations, while the producers, in fact, had footage of the inappropriate behavior. CBS defended itself by offering the following joint statement from CBS and MGM, which produces *Survivor*: "In the episode broadcast last night, several female castaways discussed the behavior of a male castaway that made them uncomfortable. During the filming of this episode, producers spoke off-camera to all the contestants still in the game, both as a group and individually, to hear any concerns and advise about appropriate boundaries" (Bradley 2019b).[1] The ultimate impact, however, was that the woman who came forward with the complaint, Kim, ended up being questioned herself, and voted off of the island.

Survivor has been referred to as a "social experiment" by its producer Mark Burnett, that somehow reflects how people will act in real life when they are put under pressure. James Poniewozick (2019) takes *Survivor* to task, though, for the way it handled the issue of sexual misconduct in the 2019 season, how it downplayed Kellee Kim's complaints, even as they finally announced that Dan Spilo was being taken off of the show. Poniewozick blames the producers for allowing Spilo to continue to touch Kim inappropriately and simply warning him against further behaving in this way. He also questions how the producers allowed the other contestants to use Kim's accusations as a way for her to think she had their support, but then voted her off the island. Viewers had to watch Spilo remain on the show, as the charges against him faded from the storyline, and they even brought in his young son at one point to further humanize him at the expense of the woman who accused him of inappropriate behavior. When they finally remove him after another incident with a crew member, they did little to support Kim, who had already been voted off the island.

In this way, *Survivor* ended up reproducing many of the complaints that the women in the #MeToo era have been raising about not having their stories be believed or taken seriously.[2] An analogy that Poniewozick raises is instructive. If someone had gotten punched by another contestant on the show, for example, that contestant would have been taken off immediately. However, in this case, a contestant is sexually harassed by another contestant and is allowed to remain for most of the season. Another analogy is if someone had a heart attack, a medic would be called immediately, as is done on the show if a contestant looks as if they are in medical distress. For the producers of *Survivor*, however, the sexual harassment of one contestant by another is viewed as less of a problem and one that can be mediated through a tribal council discussion or a verbal warning by the producers. In these ways, Poniewozick sees analogies to the CBS parent company of which it is part, who themselves were unable to deal with their own chief executive, Les Moonves, until the accusations against him could no longer be ignored.[3]

What this reveals more generally is that the themes of the #MeToo era that are being portrayed not only on reality television but in horror films speaks to the confusions that have arisen in the wake of trying to change perceptions of what it means to be sexually harassed in the workplace. Reality television shows like *Survivor* are often downplaying the impact of sexual harassment on their contestants. As we have already seen when looking at how comedies and dramas and documentaries portrayed sexual harassment in the era of #MeToo, larger cultural conversations have also at times downplayed the seriousness of these acts while, at other times, have tried to influence the conversation by portraying the devastating effects sexual harassment and assault can have on its victims. And, in terms of reality shows that center on relationships between men and women, it could be that in this era of #MeToo, these kinds of dating shows are inherently demeaning and arguably exploitative. While one strategy might be to take the genre off the air for a few years to re-work the basic premise and bring in the discussion of #MeToo, it isn't clear that these narratives can be revised to include issues of consent in a more nuanced way. In fact, reality television continues to be plagued with charges that they don't know how to handle sexual misconduct charges.

One potential strategy for viewers, then, is to adopt what Arielle Bernstein suggests, which is to draw on the lens that the #MeToo movement offers of centering the stories of the survivors and drawing on their vocabulary to question what has been viewed earlier as "basically normal" (Bernstein 2018). This orientation puts more responsibility on the viewer to make sense of these behaviors and to see how they are enmeshed in social mores from an earlier time and to challenge these media representations in light of what the #MeToo movement has brought to light. As Bernstein offered, "Perhaps one way that we can ensure that #MeToo is not just a *moment* but instead a *movement* is for viewers to pay attention to the ways that their engagement with the nuanced media that tackles these issues is part of what is necessary for things to tangibly change" (Bernstein 2018). This strategy of active engagement by viewers, combined with challenging the producers of these shows to take more responsibility for the consequences of their creative decisions, may together begin to honor the stories of survivors rather than minimize them or alternatively use them to gain higher ratings.

7

International Responses
to #MeToo

While the #MeToo movement initially began in the United States, the problem of sexual harassment and assault has been part of the international landscape of the film and television industry for decades as well. Like the Hollywood film industry, the European film industry, while viewing itself as more progressive, nevertheless has also had sexual harassment as an ongoing problem. Perhaps it is not surprising, then, that in the wake of the #MeToo movement in the United States, a French #MeToo hashtag, called #BalanceToPorc, or #SquealOnYourPig, was created as a parallel to the #MeToo hashtag. Likewise, film stars for the Cesar Awards, the French version of the Oscars, also wore a symbol to demonstrate their support of victims of abuse, a similar moment to the one at the 2018 Oscars. And, like the #MeToo movement in the United States, there was also backlash, as when French film actresses Catherine Deneuve and Brigitte Bardot called the movement "puritanical" and an attack on sexual freedom (Roxborough and Richford 2018). Though Deneuve herself eventually dialed back some of this rhetoric, this belief that the #MeToo movement was somehow intolerant was repeated by other actors and directors in the film industry in Europe. One German actress, Hanna Schygulla, for example, commented that "[w]hen I started making films [German director Rainer Werner] Fassbinder slapped me in the face and said I had to take it.... I know there is a taboo about this kind of thing now" (Roxborough and Richford 2018).

More generally, there has been a mixed reaction from the entertainment industries in Europe, who may have been less responsive to enacting changes in some countries, which may be based on cultural differences surrounding issues of race, class and gender. Audrey Clinet, a founder of a French group that helps to encourage and promote female directors, for example, believes that the civil rights movement and the women's movement in the United States has provided a context for challenging sexual harassment in the entertainment industry which Europe has not had (Roxborough and Richford 2018). Another issue may be generational differences, which may explain why older actresses, such as Deneuve and Bardot, believe that feminism is to be equated with sexual freedom rather than fighting sexual harassment. The mixed reaction, furthermore, to the #MeToo movement is not confined to France. For example, when Asia Argento, an Italian actress, accused Harvey Weinstein of raping her in 1997, when she was 21, instead of supporting her, the Italian media as well as film directors and other actors in Italy attacked Argento and instead defended Weinstein. One former MP and journalist, Renato Farina, said the assaults Argento described were nothing more than "prostitution, not rape," while another man, Vittorio Feltri (the editor-in-chief of the newspaper *Libero*), said that she should be grateful Weinstein performed oral sex on her (Roxborough and Richford 2018). When the Women's March occurred in January of 2018 in Rome, Argento invited her fellow colleagues in the film industry to join her but, when the day came, she was the only one from the Italian film industry to be represented.

Even though the European film industry considers itself more progressive than Hollywood, in terms of it being a more artistic rather than money-making community, and has espoused leftist politics on a number of issues, when it comes to the #MeToo movement, this has been far from the case. Another reason the #MeToo movement has not had the same impact on the industry in Europe as it has in the United States, is that there hasn't been a central figure like Harvey Weinstein to galvanize the movement. There wasn't a powerful figurehead that was called out, though there were some directors and producers who had engaged in sexual harassment and assault. At the same time, Europe has also had in place for a long time the idea of the auteur, who is elevated as an artist, and therefore their behavior is excused. So, when someone like Lars von Trier, Roman Polanski or Woody Allen is accused of misconduct, there is a kind of separation of the artist from the work of art that ends up condoning sexually inappropriate behavior.

Although for these reasons the #MeToo movement has not had the same impact in France and Italy as it has had in the United States, other countries in Europe have had more traction in addressing the issues raised by #MeToo. In Northern Europe, for example, in countries such as Sweden and Denmark, there is a longer tradition of gender equality compared to countries like Italy or Spain or France, where traditional cultural expectations about femininity are more deeply engrained. At the same time, however, there is also some resistance to #MeToo in these countries, despite more equality culturally between men and women. The head of the Swedish Film Institute, Anna Serner, believes that it may be due to the business model of funding for European films that is accounting for the resistance to #MeToo. Because of the public funding model in much of Europe, there is the fear that by endorsing the #MeToo goals of gender parity, it could mean that those who are currently in power in the European film industry would be pressured to make real changes. In the United States, on the other hand, there can be the rhetoric of supporting #MeToo, but there is no pressure that those in power will lose funding if they don't implement real changes as a result of #MeToo (Roxborough and Richford 2018). Despite these issues, there have still been some inroads and responses that have occurred in the film industries in Northern Europe. For example, in Finland, many women reported cases of sexual harassment in the Finnish film industry and have spoken publicly about them (Hild 2018).[1] In Germany, there have also been many accounts of harassment in the industry given by actresses. Several women have come forward and accused the film director Dieter Wedel of assault, bullying and rape in the 1980s. He had a heart attack after these allegations were published in the national newspaper *Die Zeit*, but has denied the allegations. Germany had an earlier debate around the #MeToo movement back in 2013 when the hashtag "Aufschrei" ("outcry") came out, but the #MeToo movement in the United States has reintroduced the issue for the German public.

In Spain, furthermore, while the #MeToo movement has not had its Harvey Weinstein to galvanize them, there has nevertheless been several Spanish actresses who have spoken about their experiences in the Spanish film industry. At a gala for the Spanish film awards called *Goya* in 2018, the red carpet ceremony turned into a space inspired by the #masmuieres campaign, where people from the Spanish film and TV industry spoke up and challenged the gender inequalities that are endemic to the industry in Spain. These inequalities include the sizable pay gap between men and women as well as the dearth of nominations or awards that have been received by female directors and actresses (Hild 2018). England, similarly, has had their own engagement with the #MeToo movement, both in terms of supporting it as well as seeing backlash against it. In keeping with the Time's Up campaign in the United States, however, actors and activists in the British film industry wrote a Unity Letter before the British Film industry's Bafta Awards, which challenged the industry to end sexual harassment and discrimination. Another fund was also set up, the Justice and Equality Fund, to help raise funds to end sexual harassment, and a wide array of British female actresses, including Keira Knightly, Emma Thompson and Emma Watson, have contributed to this fund (Hild 2018).

Some countries are taking active measures on the political level to address the problem of sexual harassment in their film industry in the wake of the #MeToo movement in the United States. New Zealand, for example, has created a landmark educational initiative for their entertainment industry with a curriculum created to prevent sexual harassment. The initiative was put forward at a summit sponsored by the Screen Women's Action Group (SWAG), an organization created in New Zealand in response to the global #MeToo movement.[2] The New Zealand film industry is prolific and has been responsible for some of the more innovative and award winning films internationally, as well as for producing a continuing stream of talented actors and actresses. The workshops focus on prevention, definitions and how to be respectful in the workplace. As part of the launch of this initiative, which was financed by NZ on Air (NZOA) and the New Zealand Film Commission (NJFC) the launch had a number of speakers including Prime Minister Jacinda Ardern, and *The Lord of the Rings* producer Philippa Boyens (Brzeski 2019). In this way, because of the involvement of their top elected official, as well as the New Zealand Film Commission's involvement, New Zealand is arguably helping to lead the way on how government and industry can work together to move forward on this issue. Speaking at this summit, Annabelle Sheehan, the CEO of the New Zealand Film Commission, described why this kind of initiative was necessary for the New Zealand film industry, saying "Speaking truth to power has had such huge risks and consequences for those not enjoying privilege—many have walked on eggshells for centuries. I consider conservative correctness to be the most corrosive force against freedom, creativity and equity. If so called political correctness creates a space to question words, then let it flourish" (Brzeski 2019).

There are countries outside of Europe as well, such as Israel and India, that have also experienced reactions in the wake of the #MeToo movement, though not as strong or as powerfully as what has occurred in the United States. In Israel, for example, while there is not the same kind of galvanizing figure of a serial harasser as there was in Harvey Weinstein, there has nevertheless been stories of sexual harassment and assault that have plagued the industry there. One actress, Asi Levi, described the behavior in the Israel film industry in these terms: "Our industry can't compare to Hollywood in any way, but we also have plenty of incidents of sexual harassment—in acting workshops, in acting schools, on the set. It's about 'exploiting the dream'—taking advantage of women who have dreams of fame or money. I've heard so many stories like this throughout my career—about teachers, managers and actors who took advantage of their status" (Anderman 2018). Other stories have emerged, including from Gila Almagor, the first lady of Israeli cinema, who described a violent assault she experienced in her youth when a rape scene was filmed in the movie *Queen of the Road* (1971). Other actresses have also come forward to accuse one of the most successful Israeli actors, Moshe Ivgy, of sexual harassment and assault. So, even if there hasn't been one Harvey Weinstein who has emerged in the Israeli entertainment industry, there is still a sense that there is a preponderance of this kind of inappropriate behavior that has gone unchecked. And while there is arguably a raising of awareness, including professional associations in film and television in Israel signing a convention for the prevention of sexual harassment, this hasn't directly translated into sexual harassment being entirely eliminated from the Israeli film industry. There have been some shifts in attitudes, but oftentimes it is viewed as being more careful, in case complaints might be lodged against someone. And, unfortunately, the woman may be seen as at fault if she lodges a complaint against someone (Anderman 2018).

In India, the #MeToo movement has not had a big impact on Bollywood, as their film industry is called. India has a huge film industry and is in fact that largest producer of films in the world. The industry itself is worth more than $2.28 billion (Jha 2018). Bollywood made over 1,986 films in 2017 alone, spanning 16 different languages and nine different regions in India. It is also a huge employer, with over 248,600 people working in Bollywood. Like Hollywood, however, women have fewer roles than men and are paid less than male actors. In musicals, a staple of the Bollywood film industry, they are given far fewer singing roles. The gender disparity extends not only to their earnings, but also to the fact that if you are a man, you are oftentimes protected even if you are considered a "bad boy," whereas if a woman speaks up, she is oftentimes replaced, so it has been hard for women to speak up about potential sexual misconduct for fear of losing their jobs. In September of 2018, however, a former actress, Tanushree Dutta, spoke up and described being abused by Nana Patekar, a veteran star of Bollywood, ten years earlier. After that, at least ten more individuals were accused of sexual harassment and assault, and that has included filmmakers, actors, a writer, musicians and a makeup artist (Jha 2018). Three of the men have responded by suing the women who accused them of defamation. Three other men denied that the incidents ever took place, and three have apologized to their accusers. One of the men who was accused in turn accused women who speak up of being "fat and ugly girls" who are using the publicity generated by the #MeToo movement to say they were abused as well (Jha 2018). One of the differences between the #MeToo movement in India, in contrast to the efforts in Hollywood, is that there hasn't been any substantive push for changes.

One way to account for the differences in attitude is that many people in Bollywood already knew that this kind of harassment was taking place, so it was greeted not with surprise but with indifference. For example, Shakti Kapoor, an actor who often plays rapists, was subjected to a sting operation in 2005 by news channel India TV. Kapoor reportedly offered a young reporter help in becoming a movie star if she would sleep with him, saying, "This is the way it works here…. Take it or I'll leave. I want to make love to you right now" (Jha 2018). At the time, even though the accusations against him were caught on film and were believed, Kapoor was still able to go back to work after a week. He returned to filmmaking, and the allegation had the effect of actually reviving his career as a film villain (Jha 2018). In 2005, in addition, BBC made a documentary about the film world of Mumbai and the way the casting couch culture was alive and well. The women interviewed in the film spoke about the fact that they viewed sexual exploitation as part of the price they had to pay in order to get parts in films. The actor Kapoor himself was filmed in the documentary, noting "Nobody's raping anybody here…. If the girls don't like it, they can say no and go back to where they came from" (Jha 2018). One of the cultural backdrops to these forms of abuse is that Bollywood sees itself as a kind of family as well as a film industry. The atmosphere on sets are often informal, and when abuse does occur at the hands of powerful men in the industry, it is often kept silent because if people do speak up, they are usually fired. The younger actresses are more vulnerable in these settings, while the A-list actresses, according to film journalist Janice Sequeira, "know enough to keep an entourage with them on set at all times … they rarely allow access to themselves for something to happen" (Jha 2018). And, even when women do come forward, it is hard to get their day in court. For example, 11 female filmmakers have refused to work with "proven sex offenders," but at the same time, in the Indian legal court system, there are over 30 million backlogged cases, so their cases take a very long time to be heard. In addition, there has been no legal fund established to help cover the legal fees women would incur by bringing forward charges.

China is another country where the #MeToo movement has confronted cultural as well as political barriers, including the lack of a free press.[3] While information is not readily obtainable in China, due to government restrictions on free press, there are some statistics that demonstrate that China has problems with sexual harassment, including a 2013 survey of 1,500 women compiled by the Canton Public Opinion

Research Centre, that found that 48 percent of women from ages 16 to 25 had experienced some form of sexual harassment (Zhang 2018). Writing about the entertainment industry specifically in China, Sylvia J. Martin noted that like Hollywood, the Hong Kong Media industry also has institutionalized sexism and harassment, but because of the Chinese concubine culture, which allows for and perpetuates gender inequality, it ends up implicitly sanctioning sexual harassment in the entertainment industry. As she noted, "The worry about retaliation in coming forward about sexual harassment in China would be tremendous.... In both Hollywood and Hong Kong, there is a sense, especially among older generations, that women who work in film should not be surprised by harassment—an attitude of 'what did you expect?'"(Zhang 2018). In addition, unlike the United States, China doesn't have a long history of fighting for women's rights, which might allow for women to come forward with their stories. Politically as well, because of the Cultural Revolution, women and men were exhorted to be treated equally, which has the paradoxical effect of blaming women if they come forward with accusations of sexual harassment. Another problem is that China has state-controlled media outlets, and there is censorship that works to eliminate anything posted online that might bring about social discord, including open discussions about sexual harassment. In addition, even if there were a press outlet willing to post stories of sexual harassment in the entertainment industry, there is the problem that both Hong Kong and China have very strict defamation laws, so the burden of proof on the accuser to substantiate their claims is very high.

Despite these barriers, an art house film called *Angels Wear White* (Vivan Qu, 2018) that dealt with the issue of sexual assault in China was shown at the Venice Film Festival and then released in China, even though it was an independent and low-budget film. The film explores how a migrant girl working in a motel was witness to what would eventually be the sexual abuse of two young schoolgirls by a middle-aged male companion. It then looks at what happens after there has been an allegation of abuse in a small town in China and shows how the preteen girls are subjected to victim-blaming. The film offers an indictment of a society that refuses to acknowledge the abuse that occurred. *Angels Wear White* received many nominations, including one for best director, best picture and best actress at the Taiwan Golden Horse Awards, which is the Chinese-language cinema's equivalent of the Oscars. Director Qu won the award for best director. This award may signal that change is beginning to come to China's entertainment industry, despite the cultural and political barriers that are in place. In part, this is due to the fact that, despite press and social media barriers, Chinese women are being exposed to Western ideas about sexual harassment. Ivy Zhong, female film executive and founder of Jetavana Entertainment, noted this change: "From my perspective, sexual harassment is negative and intolerable whether it's in Chinese or Western cultures ... this new generation of [Chinese] women is growing up in a multicultural world and receiving huge amounts of information each day. I believe they will create many great changes—and the film industry will be one part of the whole picture" (Zhang 2018).

Telling Stories in the Era of #MeToo

Since 2018, various projects from other countries have been announced, and some that have come to fruition, that deal with issues raised by #MeToo. The stories that were beginning to show up on American television series and films also began to appear in other countries as well. These films and plotlines were both similar in terms of themes and issues, even as the cultural milieu and attitudes may have been different from their American counterparts. For example, Oscar-nominated director Mike Figgis announced while attending the 24th Busan International Film Festival in 2019 that he was involved in a deal with South Korea's Saram Entertainment to produce three short films that are being "built thematically around the #MeToo movement" (Scott 2019).[4] The films will be using Korean actors and actresses and is being developed with Korean screenwriter Uni Hong, but will involve other Asian producers and actors.

In addition to deals that are currently in the works, there are also films that have come out during this time that deal directly with #MeToo themes. In Israel, for example, the film *Working Woman* (2019) is a story that revolves around sexual harassment. Writer-director Michal Aviad, who had begun research for this film back in 2012, reminisced that when she first began to conceive of *Working Woman*, the issue of sexual harassment was only covered in two films, *Fatal Attraction* (1987) and *Disclosure* (1994), and in both of those films, it was a female character harassing a male character (Anderson 2019). In *Working Woman*, by contrast, which was shown first at the Toronto Film Festival in 2018, the story instead looked at the harassment experienced by a married female heroine, Orna (Liron Ben-Shlush), at the hands of her supposedly kind boss, Benny (Menashe Noy). Interestingly enough, rather than make it a horrible incident of violent assault, Aviad was able to instead create a more nuanced portrayal of sexual harassment. By doing the kind of research which involves looking at actual women's testimonies, as well as her own experiences and those of her friends, Aviad offered a more subtle picture of what sexual harassment on the job can look like. She posed the following sets of questions to understand the complex dynamic involved in harassment: "What are the proximities of the bodies? What are the silences? What kind of tension is there? And while writing and thinking about it, I realized it's as much physical as psychological" (Anderson 2019). Aviad noted that Israel and the United States are both becoming more aware of sexual harassment and that there has been a rise in sexual harassment complaints in both countries. In her view, while both countries have a reactionary government in place, in Israel, because of its right-wing government that is more pro-military and has a masculinist culture, there is more tolerance of men behaving badly. That being said, the issues Aviad covers in the film are experienced by women all over the world (Anderson 2019).

The plot of *Working Woman* centers around the harassment that occurs when Orna takes a job with her former commanding officer in the Israeli army, Benny, who owns a real-estate development firm. Orna's husband is trying to make their new restaurant survive, so she takes a job with her old commanding officer to help make ends meet. The intimacy that was established in the Israeli army lends itself to a kind of familiarity that complicates matters—at one point, when Orna successfully completes a real estate deal, Benny tries to kiss her, and it isn't a collegial kind of kiss. At the same time, Benny isn't portrayed as a creepy predator, but rather as a kind of opportunistic, yet sympathetic character in his own right. In this way, Aviad makes it more complex in terms of the dynamics between the two characters. Aviad was also aware that, just as they were shooting the film, the Harvey Weinstein scandal broke, and it was clear that the film was, in a larger sense, channeling these larger cultural winds that were blowing at the time. The filmmakers, though, didn't want to turn Benny into a Harvey Weinstein kind of character. Rather, they wanted to show that Benny was surprised by the emotions stirred within himself as a middle-aged man when Orna, as a very attractive younger woman, came to work for him. They also wanted to leave it ambiguous whether he was taking advantage of her or whether he was falling for her.

At the same time, Orna feels guilty because she is worried that she needs Benny for the work and that maybe she has somehow contributed to her own vulnerability and exposure because she was trying to please him in order to keep her job. The larger point Aviad is trying to convey is that even though we may assume that sexual predators like Weinstein are how most harassers behave, it is not always so simple, and sexual harassment can take on a variety of forms that are more under the radar than they might at first seem. As Aviad pointed out, "Two of the most beloved actors in Israel were accused of sexual harassment recently … we all know these guys personally; they're not these movie villains. But you get an actor who flirts with an anonymous actress, she turns him down and then he tells the producer, 'I can't work with her.' It's this kind of stuff" (Anderson 2019). In the end, Aviad thought it would be more interesting and relevant to look at these kinds of acts of coercion and retribution because they are much more common than the egregious examples of predators like Harvey Weinstein.

In addition to movies, television shows from other countries have also touched on themes related to #MeToo. As mentioned earlier, Netflix's *Fleabag*, which won several Emmys, focuses on an English woman and her conflicts with her family, and had a few episodes that referenced male anger and sexual harassment in a humorous vein. Another show from England that came out in 2018 from the BBC One, *Press*, is, by contrast, a drama with a storyline that dramatically focuses on a sexual predator. The show is set in the current era and takes a critical look at the culture of the British newspaper industry. Created by Mike Bartlett, the show centers on the newspaper industry in England, with one paper, *The Post*, a sensational tabloid, competing with another paper called *The Herald*, formerly *The Yorkshire Herald*, which tries to uphold traditional journalistic ethics of fairness and objectivity. The editor of *The Post*, played by Ben Chaplin, was once a more responsible journalist, but is now focused on getting the most sensational stories to print, all the while encouraged by his deep-pocketed boss, George Emmerson (David Suchet), who may be modeled after Rupert Murdoch. A lot of the dialogue and plot centers on issues currently being faced in the news industry, including declining sales, journalistic integrity and regulation of the press.

The storylines are character driven as it follows the professionals who work in the industry and the conflicts they have managing their personal and professional lives, as the newspaper industry itself is subject to tremendous pressure in a global news cycle. In the third episode of the season, "Don't Take My Heart, Don't Break My Heart" (original air date September 20, 2018), the storyline revolves around reporter Holly Evans (Charlotte Riley) finding out that billionaire businessman Joshua West (Dominic Rowan) is a sexual predator. Evans is able to locate a young woman who told her that when she was 18, she was part of an internship program that the businessman sponsored, and that he groomed her and eventually pressured her to have sex with him. As several other women come forward, the businessman says that, while it may have been unethical, it wasn't illegal because all of the young women were 18 years or older. When *The Herald* tries to report on this, West gets a magistrate to sign an injunction that forces the paper to destroy all copies that were going to be on newsstands in the morning. In the next episode 4, "Magic" (original air date September 27, 2018), the editor of *The Herald* forces Evans to write an apology on behalf of the newspaper, which alienates her to the point that she goes to work for *The Post*, which ends up publishing the same story on its front page. Like the storyline in HBO's *Succession* (2018–present), the plot in *Press* that covers sexual abuse by the billionaire against multiple young women functions as a way to offer a current topic as a dramatic arc to a story that is more broadly about titanic struggles around power and the media. In *Succession*, the power plays are between the wealthy family members of a patriarch modeled along the lines of Rupert Murdoch, while in *Press*, the power struggles are more between two competing newspapers and visions of how a modern press should operate. Both series, however, used a storyline of sexual harassment to highlight these larger issues.

Another British import, Amazon's *Catastrophe* (2015–2019) has two storylines in Episode 5 of its fourth season (original air date, February 5, 2019) that explicitly reference workplace sexual harassment. The series, written by and starring Sharon Horgan and Rob Delaney, is a comedy about an Irish woman named Sharon who works as a schoolteacher and lives in England. She meets an American businessman named Rob, who works for a pharmaceutical company, while he is in London on a business trip, and they have a brief affair, which results in her getting pregnant. The series revolves around how they decide to get married, despite not knowing each other well, and how they manage their work and family life as well as their very different personalities. The show has received a lot of critical praise for the way in which it shows the truth of marriage as having a lot of rough patches and its authentic portrayal of relationships as messy and complex. In the final season of the series, Episode 4.5 (original air date, March 15, 2019) focuses on sexism in the workplace explicitly. Sharon has a new boss, the headmaster at her school. At one point he comes to her classroom when the students are not there. He proceeds to sit on her desk while she is sitting at it and leans over to talk to her, like they are colleagues who have known each other a long time, and when he leaves, she notices that there is a wet spot where he was sitting. While talking to her, he makes a comment about how the blouses she wears are lovely. While he is not being overtly sexist or inappropriate, the subtext is clearly there. When she tells Rob about the moist patch and he asks her how she will respond to this, Sharon tells him, "I dunno.... Be angry.

Nothing," which suggests that this is just something women have to live with in the workplace.

At the same time, at Rob's job, he becomes very friendly with the new company's head boss, James Cohen (Chris Noth). They "bond" together through jokes reminiscent of an old boy's club, and when Rob's immediate boss, an Indian woman named Harita (Seeta Indrani) comes in the room, they immediately stop laughing as Harita tries to make jokes along with them. They then begin to make fun of Harita's sex life, and when she tries to turn the conversation over to their work, they lose interest in the discussion and minimize her work. Cohen then gives Rob a promotion, but the promotion is to replace Harita, his own boss. The implication is that Harita is no fun and too serious, whereas he feels more comfortable with Rob. Both Sharon and Rob are dealing with workplace sexism and harassment, but coming from different sides. Whereas Sharon immediately realizes that her new boss is being sexist, Rob, on the other hand, is unaware of the implications of his boss's treatment of Harita and is offended that Sharon would identify Cohen's behavior as sexist.

This is most clearly delineated when Harita, Rob and Sharon all go out for a work dinner with Cohen. Sharon accuses Rob of "bullying" Harita, and that she's "never seen [him] be so small," but Rob feigns ignorance, saying that he and Cohen were just joking around and that Harita is a strong woman who could take care of herself. Later on, after Rob and Sharon confront each other about this, Sharon then tells her own boss in a light hearted way that he shouldn't make comments about what she is wearing and he shouldn't be sitting on her desk. The headmaster doesn't say anything, but it is clear he is angry with her and, later, Sharon says that maybe it is time for her to look for another job because she has angered her boss. Even if she stays, there will always be awkwardness between her and the headmaster, and because he is her boss, she doesn't have much leverage to make the relationship better. At Rob's job, in the meantime, he decides it would be unethical to take Harita's job, so he turns it down. However, rather than be proud and happy that her husband wouldn't betray Harita, Sharon gets upset because she feels like since her own job may be precarious, their family needs the promotion and pay Rob would have been able to get if he accepted the new job, or as Sharon tells Rob, "I just don't think we can afford to be idealistic ... with our mortgage and credit-card debt, and you know, you said you wanted your teeth whitened."

For Horgan, along with Delaney, these plotlines of sexism in the workplace were more organic to the storylines in Season 4. As she noted about the squirt on the desk, which opened up the story on sexual harassment, "I'm not sure what it was that brought the idea to us, whether we thought of the squirt on the desk before we thought about what we were going to do with it.... Once we realized what was there, what we had and once we realized we had a concurrent Rob work storyline that is also its own sort of how *not* to behave, we thought 'let's make a virtue of it and let's have Rob and Sharon not making the right choices either.' You know, knowing what the right thing to do is and knowing how to do it are two completely different things" (Sarachan 2019). In this way, while it was not an overt desire to make a statement about the #MeToo movement, Horgan and Delaney drew on what was occurring in the larger culture around sexism in the workplace to illustrate the tensions their characters exhibited when confronted with making the right choices around moral dilemmas.

In some ways, the impact of #MeToo on foreign films and television can be measured not only in specific storylines that include #MeToo themes but also in terms of the kinds of characters that are emerging in recent narratives, specifically, in presenting women who are actively dominating the story, rather than being passive victims. In the Oscar nominations for the 2020 international features, for example, many featured female characters who were the main protagonists in the film. One of the films, *Queen of Hearts* (2019), is from Denmark and is about a 40-something-year-old lawyer who tries to seduce her teenage stepson. The story is inspired by true events and tries to understand why the woman would do such a thing. There is another film from Canada, *Antigone* (2019), that is made by a woman, Sophie Deraspe, and tries to put a modern spin on the Greek tragedy about a woman who avenges her brother. In another film, *Instinct* (2019), from Holland, a female prison therapist (Carice van Houten) has an affair with a serial rapist (Marwan Kenzari) who is an inmate in the prison. In the case of *Queen of Hearts*, the director is a woman, May el-Toukhy, and when asked why she would create such a dark female character, she noted "if we want to have more female leading characters, we have to create women that are nuanced, subtle and sometimes dark" (Roxborough 2019).

This focus on having leading female characters who are multi-dimensional can especially be seen in the Russian film *Beanpole* (Dylda, 2019). The film, the second film from director Kantemir Balagov, centers on two Soviet Army women who were sent to fight with the men on the front lines of World War II and follows their lives in Leningrad during the first months after the war. One of the women, Iya (Viktoria Miroshnichenko), nicknamed Beanpole, is discharged from the army because she is suffering from shell shock. At the same time, her best friend Masha (Vasilisa Perelygina) has to stay at the front, but she is also traumatized by the war. While both women are in shock, we see that Iya is subject to episodes where she is in a kind of trance, the result of a concussion she suffered while in the war. During one of these trances, she smothers a small child who is in her care. It turns out that the child belongs to Masha, and Iya had been caring for him while Masha remained on the front lines. Upon learning that Iya killed her child, Masha is overcome with grief. She tries to get pregnant again by coercing a young man to have sex with her. The story unfolds as the two women treat each other abusively, even though they love each other, because they are both so damaged by the struggles they endured during the war and after. Balagov said he based his film on a 1985 oral history of Soviet women, compiled by Svetlana Alexievic, who had lived through the war and were "surrounded by death" (Roxborough 2019).

What makes all of these stories arguably a reflection of some of the changes brought about in the #MeToo era is that storylines are explored from a woman's perspective. Speaking about her work in *Instinct*, for example, director Halina Reijn noted that she feels part of the #MeToo movement and that the movement has "brought about a lot of positive changes, also in the workplace" but that she doesn't think it should be about placing restrictions on what kinds of stories about women get told because "as an artist, we should never think in good and evil, never think in black and white. We should always confront, be courageous and go into the spaces we are so afraid of" (Roxborough 2019). At the same time, Reijn wanted to make sure that the film was told from a woman's perspective and that there would be absolutely no female nudity in the film, even as there are explicit sex scenes. She said that she was tired of objectifying women's bodies in film and stage, and that for a change, the male body can be objectified (Roxborough 2019).

At the same time, El-Toukhy, director of *Queen of Hearts*, also cited the #MeToo movement as helping her to ask the hard question of whether her film could be a reflection of the movement. She was concerned that people would somehow view her film as anti-female because it portrays a woman in a negative light as a seducer. She then described how, when the film was shown at the Sundance Film Festival, a man got up during the question and answer sessions and said, "Now I understand what #MeToo is about. Now I get it" (Roxborough 2019) For El-Toukhy, this helped her clarify her own intentions in making the film, noting "And that's what we wanted do with the film. To show that the #MeToo movement is not just about gender. It's about power" (Roxborough 2019). Though the #MeToo movement originated in the United States, directors like El-Toukhy feel that their films wouldn't have been made in the U.S. because the American system of filmmaking is based on box office sales, which means that Hollywood is more averse to making films that would potentially alienate or scare an audience. For these foreign filmmakers, who work in Scandanavia and other European countries, and whose films are financed by public funds, there is more latitude in the kinds of stories that can get told. In these ways, one of the lessons from the impact of the #MeToo movement is that it is in the complexity of the portrayals of women characters that the measure of the movement can, in part, be taken. To have stronger and more complex portrayals of women onscreen and to have their stories told may be a better way to understand the impact of the movement in the coming years. As Reijn noted, "For me, that is the whole thing. That is feminism in cinema.... To explore female identity in all its complexity. So don't make her into a hero. That would not be fair and that would not be real feminism for me. Because what is a hero? It's a flat character" (Roxborough 2019).

Conclusion: "Freedom to Bother" vs. Substantive Change

To summarize, while each country has had different reactions to the issues of gender inequality and sexual harassment raised in the wake of the #MeToo movement, it is clear that there has been some movement to try and open up the kinds of stories being created in international film and television industries. There are some attempts to come to terms with the need to change the gender dynamics of these industries, as witnessed by the 2018 Cannes Film Festival, where the organizers created a sexual harassment hotline and handed out tote bags that included fliers with warnings that misconduct would result in a huge fine and imprisonment. At the same time, if we take the same country, France, there has also been a counter response which can be seen in such examples as the French public reaction to an allegation by the actress Sand Van Roy, that the filmmaker Luc Bresson had raped her more than once, as well as the publicly signed letter by women in the film industry saying that the #MeToo movement has gone too far. In both cases, there was a reaction that challenged the #MeToo movement as going against French culture and its assumptions about relations between the sexes. This controversial debate about the #MeToo movement has continued in France unabated but, at the same time, accusations against filmmakers continue to arise.

The actress Adele Haenel, for example, said in 2019 that she had been abused as a child by a movie director, and she became one of the first actresses in France to speak publicly about her experience and the abuse in the country's film industry (Peltier 2020). The director, Christophe Ruggia, who Haenel said began sexually harassing her when she was 12 years old, denied the accusations, but was charged in January 2020 with sexual assault on a minor under fifteen. Since then, another accusation against Roman Polanski has come out from Valentine Monnier, a photographer who said Polanski raped her in 1975 when she was 18 years old (Peltier 2020). For Haenel, the fact that it has taken so long for women to feel like they could come forward signals a lax culture that treats claims about sexual harassment as somehow adopting a puritanical and American attitude rather than in keeping with France's cultural acceptance of the blurring of lines around work and sexuality. As Haenel noted, "here is a #MeToo paradox in France: It is one of the countries where the movement was the most closely followed on social media, but from a political perspective and in cultural spheres, France has completely missed the boat ... many artists blurred, or wanted to blur, the distinction between sexual behavior and abuse. The debate was centered on the question of [men's] 'freedom to bother,' and on feminists' purported puritanism. But sexual abuse is abuse, not libertine behavior. People are talking about it, though, and #MeToo has left its mark. France is boiling over with questions about it" (Peltier 2020).

In France, then, and in many other countries, while the entertainment industries have similar challenges in terms of the entrenched attitudes about acceptable behavior and whether women have been victimized, changes have begun to occur in how sexual harassment is thought about, which mirrors changes happening in the larger societies of these countries. When women do come forward, however, they still often face a judicial system that presents large challenges in getting their cases heard. Even as there exists a concerted effort to change the system, there are still many barriers that remain, and entrenched attitudes are hard to dislodge. France's shifting cultural attitudes demonstrates these tensions. For, just as they had one awards festival that included awareness of sexual harassment, in the next instance, at the Cesars, Roman Polanski received award nominations, which for Haenel is equivalent to "spitting in the face of all victims. It means raping women isn't that bad" (Peltier 2020).

At the same time, when women challenged these nominations, other men and women objected that it was a form of censorship to not watch Polanski's *An Officer and a Gentleman.* Still, the changes the #MeToo movement called for can be seen in many countries, including France, and may reflect a larger cultural awareness beginning to take hold in these societies. France, for one, has seen a large increase in

Pedophile genius: Roman Polanski on the set of *The Ghost Writer* (Summit Entertainment, 2010).

reported sexual crimes and in the number of women coming forward to file complaints. A recent report which catalogued these changes found that the willingness of victims to come forward "had been 'amplified' by the accusations of abuse against the Hollywood producer Harvey Weinstein and by the associated #MeToo movement" (Breeden 2019a). In these ways, then, the international response to the #MeToo movement is still being written and is reflected in the courage of women from across all industries and walks of life to come forward, even as their societies react with uneven and mixed responses.

At the same time, as we have also seen, the stories being told on screen have begun to fully question not only the forms that sexual harassment can take, but the very institution of patriarchy and the assumptions around power relations that exist in these institutions. In the French film *Portrait of a Lady on Fire* (2020), this perspective is revealed through an exploration of a love story between two women. The film is a story that takes place at the end of the 18th century. A painter named Marianne (Nicole Merlant) is commissioned to paint a portrait of another young woman, Heloise (Adele Haenel), but has to paint her in secret without Heloise knowing that this is being done (Clements 2020). The director, Celine Sciamma, wanted to create a love story that didn't fetishize a woman and in so doing, as Merlant notes, ends up "represent[ing] a cinematic experience that is no longer ruled by the male gaze" (Clements 2020). For an actress like Merlant, the way Sciamma told the story allowed for a number of issues to be talked about, including abortion, from a female's perspective and, for this reason, it ends up being an important byproduct of the #MeToo era. As she observed, "The place of women in society, in culture, and in art is something we're really thinking about … we realize that, in a way, we've erased the history of women because we only know what has been shown to us through one particular lens. I think that's why the film has had such a big impact on people because they've been waiting for a vision like this" (Clements 2020). Hopefully, more of these stories will get told as a result of the opening up opportunities in the #MeToo era for female filmmakers in a variety of countries.

Conclusion

Will #MeToo Ultimately Make a Difference?

On February 1, 1978, Roman Polanski fled the United States after having pled guilty to having unlawful sex with a minor, 13-year-old Samantha Gailey (now Geimer), in March 1977, when he himself was 43 years old. Polanski had taken her to Jack Nicholson's house ostensibly to take photos of her for a magazine, at which point he gave her champagne and Quaaludes and then forcibly entered her, anally. Since that time, several more women have come forward and similarly asserted that Polanski had sex with them, forcibly, when they were under the age of 18. Before the #MeToo movement, the view in Hollywood and Europe (Polanski fled to France in that period and lived ever since) was that Americans were unforgiving of him because of the country's puritanical reputation, with an overly conservative approach to sex. For many actors, however, far from agreeing with the verdict of his guilt by the U.S. court or the court of opinion in the United States, instead found it desirable to work with him. Such diverse actors as Sigorney Weaver, Ewan McGregor, Kate Winslet, Johnny Depp and Harrison Ford have all starred in his films (Freeman 2018a). Though the facts have never changed from what the initial testimony revealed, these and other actors justified their working with Polanski as well as Woody Allen, another director accused of sexual assault against a minor, because of their "extraordinary reputation," or as Kate Winslet offered, "Having thought it all through, you put it to one side and just work with the person. Woody Allen is an incredible director. So is Roman Polanski. I had an extraordinary working experience with both of those men, and that's the truth" (Freeman 2018a).

In the wake of the Harvey Weinstein scandal, and the #MeToo movement that followed, it is helpful to contrast the shifting responses to Polanski from comments like these made in earlier times to now. For, as Hadley Freeman has pointed out, these actors had knowingly

151

worked with Polanski before the #MeToo movement, had known he was a convicted sexual assaulter, and thus "tacitly approved" of rape (Freeman 2018a). In the case of Winslet, she ultimately changed her position and admitted to having "bitter regrets" and said, "I have at times made poor decisions to work with the individuals whom I wish I had not. Sexual abuse is a crime, it lies within all of us to listen to the smallest of voices" (Freeman 2018a). In fact, she, and everyone else in the entertainment industry, knew about the accusations against Polanski and Woody Allen for decades. So, what has changed? For, before this time, these actors and producers exonerated Polanski, including Harvey Weinstein himself (who said in an open letter that "whatever you think of his so-called crime, Polanski has served his time") as well as Whoopi Goldberg (who said Polanski hadn't committed "rape-rape") to Johnny Depp (himself accused of domestic violence against his then wife Amber Heard), who said about Polanski "very clearly, and he's proven … he's not a predator" (Freeman 2018a).

What has arguably changed is that the Weinstein scandal and the #MeToo movement have altered how sexual harassment and sexual assault, and those who have been accused of it, are understood. In the case of Weinstein and Allen, they are no longer given a "free pass." Unfortunately, Roman Polanski is still considered a maligned artist by many, despite the fact that, in an interview he gave after he fled the U.S., he said, "Judges want to fuck young girls. Juries want to fuck young girls— everyone wants to fuck young girls!" clearly indicating that he saw no problem with his actions that led him to flee the United States (Freeman 2018a). If there has been some movement, then, on the part of actors who had previously condoned the behavior of people who committed sexual harassment in the entertainment industry, what of the audiences themselves? On May 10–14, 2018, a poll was conducted among 2,202 U.S. adults where respondents were asked whether they would watch a show where allegations of sexual harassment against a star had been made. Of the 20 entertainers who had been listed in the survey, only two entertainers, Kevin Spacey and Louis C.K., were viewed as "unwatchable" given the accusations against them. Kevin Spacey had a higher negative impact in terms of whether people would watch his work (46 percent) versus not having an impact on whether they would watch him (36 percent) (Piacenza 2018). What these statistics reveal is that, while the #MeToo movement tried to call attention to the sexual harassment that exists in the entertainment industry, the impact on viewership habits may be limited. And, there is some question as to whether, even in the case of Spacey or Louis C.K., the negative impact will be lasting.

The career of Mel Gibson may be instructive in this instance. While he was once considered "box office poison" for his anti–Semitic and racist comments, he was able to reclaim his reputation when he earned an Oscar nomination in 2016 for *Hacksaw Ridge* (2016), which earned over $175 million worldwide. Another example is the case of Jeffrey Tambor, who was accused of sexual harassment on the set of *Transparent* (Amazon Prime Video, 2014–2019), and went on to work on *Arrested Development* (Netflix, 2018–2019). Seventy-one percent of viewers believed that Netflix's decision to keep him on the show wouldn't make a difference in terms of whether they would view the show or not (Piacenza 2018). At the same time, 72 percent of respondents said the same thing about Amazon Studios' choice of removing Tambor from the show *Transparent*. Whether he was kept on a show or removed from a show, in other words, had the same impact on viewers, with the majority saying it wouldn't make a difference. For these respondents, the only actor who had a lot of name recognition in terms of allegations was Kevin Spacey. One possible reason for these responses is that for the entertainment industry and the actors who work in it, the sheer number of allegations against so many different people may account for why any individual allegation against an actor has a hard time of registering with audiences (Piacenza 2018).

Actors' and Actresses' Views of Perceived Change

Besides looking at audience reactions, one of the ways that change can also be measured is to focus on the people who work in Hollywood to see what they believe has been the impact of the #MeToo movement. While Kate Winslet represents one shift in attitudes in terms of whether they would continue to work with known predators, the larger question of whether the industry itself has changed for the better is an open question. Laura Dern, an actress who has been in Hollywood for a generation and who won the 2020 Golden Globe Award for Supporting Actress for her role in *Marriage Story* (Netflix, 2019), believes there have been changes as a result of the movement, specifically in terms of the way young actors from the new generation are being treated (Erbland 2018). She noted that there is now a concerted attempt through unions such as the Screen Actors Guild to revise the code of conduct that would support and protect actors in the audition process, which had been one of the places where harassment was likely to occur. She was also encouraged by the work being done by younger actors in particular who are not waiting for the older generation to enact changes. In her own life, the movement had an impact as well, in terms of the permission she finally gave herself to understand her own story. In terms of the industry, she was also encouraged that there is now a push for better safety measures on sets, both in terms of creating boundaries around sexuality as well as physical safety (Erband 2018). Many other women working in the entertainment industry have similarly described their sense that there is a lot more listening going on, and more awareness of how people are speaking up about things. Actresses like Patricia Clarkson, for example, who starred in *Sharp Objects* (HBO, 2018) and has been acting for decades in Hollywood, talked about how the tone has changed in meetings. As she noted, "I would say the way certain male directors have spoken to me is different. They're just quieter. I feel it, and I maybe want to believe it, but I feel a difference. I feel a tectonic shift has taken place in Hollywood for the first time" (Vulture 2018). For others, however, the fact that people are speaking up doesn't mean that things have really changed in terms of the power structure. Elizabeth Reaser, an actress who started in *The Haunting of Hill House* (Netflix, 2018), gave an example of working on a set where most of the people are still men, noting that even if there are more actresses working in Hollywood, she wasn't optimistic that this somehow signaled wholesale changes in the industry (Vulture 2018).

It is also interesting to see whether there are differences as time goes on, after the initial outpouring of publicity around #MeToo. For example, several actresses described feeling a change in tone on the sets that they are working on and in the emails they get, to even the questions that are asked of them on the red carpet. One actress, Kerry Bishe, who has starred in *The Romanoffs* (Amazon Prime Video, 2018) and *Halt and Catch Fire* (AMC, 2014–2017), describes getting an email from Andy Bobrow, a showrunner she used to work with and who she did a Fox pilot with. He apologized to her for how he behaved during an audition, telling her to smile more, or as she described it, "smile, all the time. How are we going to know that you're friendly or nice so that we can root for you if you're not smiling? And you work with it, you roll with it, and you fucking smile!" (Vulture 2018). Another actress, Emily Meade from *The Deuce* (HBO, 2017–2019), talked about having been in the entertainment industry since she was 18 years old. In her very first film, she was required to be in a sex scene. She said she has been put in very sexualized roles throughout her career and that she lost certain roles because she objected to the fact that she thought the characters were sexualized for no reason (Vulture 2018). Meade didn't think she had any control over the situation and was upset that she lost opportunities because she didn't want to rely on the sexualization of the female characters to get acting roles. After the first year of #MeToo, she found that her work on *The Deuce* for HBO was more open because they now had an intimacy coordinator for the sex scenes, and that moving forward the impact of #MeToo will not only be in terms of bringing attention to the problem but finding solutions to it (Vulture 2018).

One Step Back?

While there are many actresses like Dern and Clarkson working in Hollywood today who do believe that some changes have occurred in the atmosphere and the culture of the industry, the fact that so many men who had been accused of sexual misconduct but have gone back to work also suggests that the changes are by no means permanent. As well, the fact that certain woman-themed shows have been canceled while other male-themed shows are allowed to continue, even though both have relatively few viewers, indicates that the culture of Hollywood has not shifted ground in any substantive way. For example, an AMC show called *Dietland* (February 2018), created by Marti Noxon (creator of *Buffy the Vampire Slayer* [WB, 1997–2003 UPN] as well as *Girlfriends' Guide to Divorce* [Bravo, 2014–2018]), was about a group of women who kidnap and murder men who have hurt women. The theme of the show resonated with the anger generated as a result of the #MeToo movement. Even though the show won critical praise, it also generated low ratings, with only about a half a million people watching its season finale and by September of 2018, it was canceled. At the same time, *Lodge 49* (AMC, 2018–2019), which was about a surfer who bonds with men in a fraternal lodge after his father's death, also had drawn only about 600,000 viewers, yet was renewed over the show about women's anger (Gilbert 2018a).

While Hollywood did take some positive steps in terms of policies that would address workplace harassment, including putting in new rules and training programs and ousting the most egregious of harassers, the areas that are more gray and nuanced are harder to identify as having been substantially altered. This includes such issues as who ends up being promoted, which shows get renewed, which films are shown at film festivals, and so on. In 2018, for example, the advocacy group "Women and Hollywood" published statistics that showed only 27 percent of those working behind the scenes in television in Hollywood (including directors, writers, editors, producers, directors of photography, etc.) were women (Gilbert 2018a). This figure, which is lower than the figure for the previous year, was also mirrored in the number of speaking characters on television, with women being only 40 percent there, and which again was lower than the year before. And, in the 2020 Golden Globe Award nominations, there were no female directors nominated at all.

In terms of the question of representation, then, there has also been less progress, both in terms of who is allowed to be behind the scenes, and also which projects get greenlighted. Such shows as *The Romanoffs* (Amazon Prime Video, 2018), *Escape at Dannemora* (Showtime, 2018), *The Good Cop* (Netflix, 2018), *Into the Dark* (Hulu, 2018–present), *The Little Drummer Girl* (BBC One, 2018) and *The Kominsky Method* (Netflix, 2018–present), all of which are led by men and mostly about men, were all in the fall 2018 lineup. And many of the individuals called out in the wake of the #MeToo movement have been able to return to their careers. In addition, as we have seen, after the Weinstein scandal broke in October 2017, within some months a kind of backlash began where it was thought the movement had gone too far.

In fact, the backlash against #MeToo has been very public and is tied to a larger narrative that the individuals accused of misconduct have been treated as if they are guilty without a trial. Oftentimes, while there is a general recognition that it has been good for Hollywood and the larger culture to come to terms with sexual harassment and assault, there is oftentimes then a plea to be more sensitive to the people who have been accused, as if they are the victims of a movement that has gone too far. This can be seen in such examples as Andrew Sullivan, a columnist for *New York Magazine* who likened #MeToo to a kind of sexual McCarthyism, or the journalist Masha Gessen, who compared the movement to sex panics (Garber 2018). Oftentimes, the defense of men who have been accused of harassment is that the men had done nothing wrong other than being somehow politically "incorrect" or casually lazy, rather than predatory. Bill Maher, for example, has accused the movement of creating a situation where, to keep things 100 percent safe for women, would in effect make a "police state" for men (Wilstein 2018). The operative principle here is that the empathy is focused on the men who have been somehow wrongly accused rather than the women who have made the accusation, and this "selective empathy" has resulted in a kind of equivocation in many cases that has allowed the men to not only claim their innocence, but to make themselves out to be the victims.

And What of the Accused? What Happens to Them?

In light of the backlash that has occurred, there has been an inconsistent response by the entertainment industry which has been justified, in turn, by the perception that the movement has gone too far. This inconsistency is demonstrated no more clearly than with the question of how to move forward after someone is accused of sexual harassment or assault. In some cases, these individuals are fired from projects, only to return to the industry at a later point. In others, they make public apologies or public denunciations of the accusers as engaging in a "witch hunt." And, perhaps most tellingly, in many of these cases, the voices of the women in the industry only come to light if they themselves have a high degree of power and influence or if people come forward in large enough numbers to tell their stories. As the culture is coming to terms with what it means to live in a post–#MeToo world, the women who were actually harmed as a result of sexual harassment in Hollywood are also trying to find justice and compensation. The victims of Harvey Weinstein, for example, are currently negotiating the payout of money as a result of the tentative $25 million settlement with the Hollywood producer. There are over 30 victims in the lawsuit, and they include a wide variety of women who were either former employees or actresses and are from not only the United States, but also Britain, Ireland and Canada (Gupta 2019).[1]

Jodi Kantor, one of the original authors of the exposé on Weinstein for the *New York Times*, explained the dilemma in these terms: "What's the right recompense for these kinds of allegations? It's a hard question. Consider that a single one of [former Fox News host] Bill O'Reilly's accusers was paid $32 million, while in this tentative civil settlement, dozens or more of Weinstein's accusers will split the $25 million" (Gupta 2019).[2] More generally, Hollywood continues to employ men who have had accusations of sexual harassment lodged against them, and they continue to greenlight films and television shows centered on themes about men far more often than about women. This is happening despite the fact that there are concerted efforts to eliminate harassment in the workplace in Hollywood and beyond, including important work done by SAG-AFTRA, the actors' union, around where auditions can be held as well as offering counseling services to help their members deal with harassment if it happens to them. Time's Up, in addition, was able to help over 3,500 women and men get legal representation for their claims of harassment in 2018 alone. And, as noted earlier, there are now attempts to get protection for actors who have to appear naked on camera by having "intimacy coordinators."

For some, these contradictory aspects of Hollywood, with some attempts to address harassment, such as the replacement of some of the worst offenders including Les Moonves of CBS or Roy Price of Amazon Studios, but at the same time allowing others to remain, means that there is not a unified response to the problem of harassment and gender equity in Hollywood. This has sent a dual message—that you shouldn't abuse your power in Hollywood, but even if you do, you could still have a career there. Sophie Gilbert has pointed out that one of the reasons for this is that in Hollywood, with all of its stories about fathers and sons, buddy cops and prison breaks, and male grief, people may be more likely to forgive the men in their ranks and give them a second chance (Gilbert 2018a). This leads to the problem that, for all the potential of the #MeToo movement, there is still a long way to go to fundamentally alter the culture and practices of the people producing those stories.

One of the questions moving forward is what kind of response is the best one to deal with the people who have been accused of sexual misconduct and who want to return to their former professions. In some cases, the offenders actually faced criminal charges, as is the case with Harvey Weinstein. In most cases, however, there hasn't been any kind of formal process of charging people, and the charges themselves may stem from acts committed several years ago. Fatima Goss, president of the National Women's Law Center, which is in charge of administering the Time's Up Legal Defense Fund, has found that while the harassers want to return to their jobs and are oftentimes able to do so, there is not a lot being done to help the victims who have accused them. She noted that companies who had hired the individuals accused of sexual harassment need to make sure they deal with them and the women who were affected by them, as well as the environment which allowed these kinds of behaviors to take place (Green and Sakoui 2019).

Goss also noted that they should expect to pay some kind of reparations to the victims. In some cases, the harassers are financially well off enough to be able to set up their own businesses, as happened in the case of Les Moonves, who worked for CBS but was able to set up another business funded by CBS after he was initially fired (Green and Sakoui 2019). In other cases, it can be up to the public to decide when and how they are willing to forgive the individuals accused of harassment, as happened with Louis C.K. and Aziz Ansari, both of whom are back to performing for the public. In other cases, the act of rehabilitating these individuals would include some kind of compensation for their victims, especially for the work they lost as a result of the harassers derailing their careers. Another idea is the critical element of acknowledging what happened and apologizing to the victims.

Impact on Those Who Came Forward

Another way to measure the impact of #MeToo on Hollywood is to think about the way the victims of sexual harassment have been treated. There are risks for speaking up that can cause further damage to careers already negatively impacted by the harassment, though there may be less risk for someone who is famous, wealthy or privileged in other ways. Rebecca Traister (2019) spoke to many women both inside and outside of the entertainment industry who came forward to try to understand the repercussions on their professional lives. She found that their critics saw them as part of a mob of angry women, where others greeted them with a sense of "thrilling sisterhood" (Traister 2019). For many of these women, there was a sense of loneliness and isolation and a reluctance to talk about the harassment they had kept silent about for months or years. Unless they were already famous, many of the women were relatively anonymous and forgotten about quickly, and few if any of them received any kind of personal gain from coming forward.

Traister makes the point that much of the conversation in the entertainment industry and beyond has been about the redemption of the harassers and how they should get their jobs back, but little if any conversation has concerned the difficulties for the victims who came forward and who have further risked their careers by going public. And far from taking pleasure in calling out the behavior of these harassers, many of the women had a range of emotions for the harasser, ironically enough, from sadness to guilt to pain. Describing the public's own reactions to these harassers, Traister observed that "[i]It may be fucking twisted, but it's what we do: We crane our necks to see the wreckage of powerful male careers without even bothering to wonder about the women whose lives and careers those men damaged. Because it was the men who were powerful, some of them already familiar to us, and because they were men whom we have been encouraged to view as fully human, we are led, often unconsciously, to be more fascinated by their stories, to understand them as complex and nuanced and interesting characters, even in their villainy. A scrap of ambiguity entrances us in powerful men, while we find less dramatic or interesting the complexities and internal contradictions of those who stepped forward against them" (Traister 2019). Traister found that, by and large, the women who came forward paid a professional and personal price, and she ultimately likened them to how another colleague characterized them—as "foot soldiers" who were sacrificed as part of a larger war.

Changes in Culture and Business?

For these women, then, it is important to ask whether there have been real changes in the culture and business of Hollywood, and beyond that, if those changes are worth the sacrifices they made. The #MeToo movement has had an undeniable impact on the culture and business practices of Hollywood. This could be seen when Kevin Spacey was removed from the film *All the Money in the World* (2017), which went on to become an Oscar contender. Christopher Plummer took his place, but at a cost of $10 million more for the filmmakers. After the film was released, in addition, it emerged that Mark Wahlberg was paid an additional $1.5 million to reshoot his scenes with Spacey while his co-lead Michelle Williams was paid less than $1,000. This in turn created an uproar at which time Wahlberg ended up donating his additional salary to Time's Up Legal Defense Fund (Zacharek 2018).

Another change #MeToo has brought can be seen in some film festivals. In 2017, only 8 percent of the top-grossing films at the Sundance Film Festival were directed by women. In 2018, by contrast, 37 percent of the festival's films were directed by women (Watercutter 2018). One of the films was *See Allred* (2018), a documentary about Sheila Allred, the women's rights attorney who has been involved in several sexual harassment cases. Another movie was about Supreme Court Justice Ruth Bader Ginsburg. A third film was about queer women who were sent to gay conversion camps. During the festival there were several panels and discussions that covered the topic of women in Hollywood as well. On one panel, "Women Breaking Barriers," Octavia Spencer spoke directly about the ways in which the #MeToo movement has inaugurated a new movement to empower women in Hollywood. For some, the Sundance Festival itself was seen as having the potential for making lasting cultural changes in Hollywood, including giving more opportunities for women filmmakers to have their films shown. Other participants, like Jennifer Fox, who directed *The Tale* (HBO, 2018), a film about dealing with the effects of childhood sexual abuse based on her own life experiences, talked about the need for more funding for female-fronted projects, and more generally, tied the question of funding and representation to issues like sexual harassment (Watercutter 2018). Like unequal wages, a lack of funding or opportunities have kept women back in Hollywood and resulted in fewer films told from a woman's view, so women's stories are not heard. Even as these stories begin to make their way into film festivals, there remains the challenge of finding a wider distribution that would allow audiences beyond Sundance to see these films.

The 2018 Oscar nominations also revealed that Hollywood as an industry has evolved somewhat in its representation. Greta Gerwig, who directed the film *Lady Bird* (2017), was one of only five women to be nominated for Best Director since the awards were given.[3]

Rachel Morrison became the first female director of photography to be nominated for Best Cinematography for her work on *Mudbound* (2017). Despite these gains, there is still a sense that not enough has been done to level the playing field in Hollywood. The business as a whole still doesn't have a single reporting system for sexual harassment. Long-time filmmaker Nicole Holofcener feels that this is part of a larger problem in Hollywood where not enough has been done to reduce the inequality in hiring and protecting women and people of color (Coyle 2018). As she noted, "It feels like we're moving in the right direction, but women and minorities are such a tiny percentage of this industry.... I open up my Director's Guild magazine, and it has films that the DGA is screening and sometimes there's not one woman, not one black person. They are all white male directors and my jaw is on the floor. I think: How can this still be?" (Coyle 2018). That's why, even as the tone has shifted at film festivals or awards shows like the Golden Globes and the Oscars, where #MeToo sentiments could be heard in the speeches and on the red carpets, the actual amount of substantive changes are harder to gauge. And for many, the sense is that it won't be fundamentally different until there is more equity in the industry. Kristen Schaffer, the executive director for the advocacy group Women in Film, believes that parity will be the ultimate way of eliminating harassment. As she noted, "The more women we have in leadership positions, the less likely the incidents of harassment. So we have a lot of work to do on that front.... We've been living in a sexist, racist society for hundreds of thousands of years; we're not going to undo it in a year" (Coyle 2018).

One of the main ways the entertainment industry is trying to rectify many of the issues raised by the #MeToo movement is to develop a survey on systemic bias and abuse in the industry. This survey is being sponsored by a coalition of Hollywood's largest companies dubbed the Hollywood Commission. It is being led by Anita Hill, who herself became famous in 1991 when she had to testify in Senate confirmation hearings about having been sexually harassed by then–Supreme Court Justice nominee Clarence Thomas. The commission consists of people from movie studios, record labels and talent agencies, and will survey anyone who has worked in these industries through an anonymous poll. Drawing on a variety of people from these fields, participants will be asked to answer questions about their experience working in the entertainment industry. Speaking to the need for this kind of survey, Hill commented that "[d]ue to the heroic and brave work of many, we all now know there are serious problems of harassment, bias and mistreatment of others in Hollywood" (Shaw 2019).[4]

Another way to gauge the impact of the #MeToo movement and how it can go forward is to look at some of the women who were at the forefront of the movement. In November 2019, the comedienne Samantha Bee hosted an episode of her show *Full Frontal* (TBS, 2016–present), where she invited Tarana Burke, Megan Twohey, Jodi Kantor, Chanel Miller and June Barrett to talk about the #MeToo movement as it reached its two-year anniversary (Schaffstal 2019). Bee described this anniversary as a "trash fire that's still burning Matt Lauer's pubes," and proceeded to host a Thanksgiving dinner with the women to talk about where to go next. Tarana Burke noted that a lot of the indivdiuals from the media that contact her now are more focused on asking her how the men who were accused could find the "road back" (Schaffstal 2019). At the same time, while many of the women who lodged the accusations are no longer in the news, talking about the men who were accused is still media-worthy. Kantor and Twohey, the reporters who broke the story about Weinstein in the *New York Times*, reflected that they had no idea what a tremendous impact their story would have. Chanel Miller, the young woman who had initially remained anonymous about being attacked by a student at Stanford, felt that finally revealing her identity had allowed her to "transcend" the attack (Schaffstal 2019). Barrett, who had also been raped, described the same feeling that by coming forward with her story, she not only began to heal herself but felt that it was an important step in helping other women to come forward. All of the women mentioned that federal sexual harassment laws are still not strong enough, and that there is still the phenomenon of being paid by men to remain quiet about having been assaulted. Tarana Burke then explained that she had started another hashtag, #MeTooVoter, to help elect a candidate for president who will support people who have experienced sexual violence.

Nudity in Film and Television

Another way that actresses have responded to the #MeToo movement is to begin to speak out against doing nude or other sexual scenes. For example, Emilia Clarke, who was on HBO's *Game of Thrones* (2011–2019), was initially a relatively unknown young actress when she got her break playing the character Daenerys Targaryen on *Game of Thrones*. She described her reaction to seeing how many nude and sex scenes there were in the script for the show, but felt like she couldn't say no to doing them, and would end up crying in the bathroom on the set before each of those scenes. As a new actress who hadn't been in the industry long, she found herself naked in front of an entire set, and it was only because of her co-star, Jason Momoa, who told her that it wasn't okay to be treated like this, that she felt like she could speak up (Fry Schultz 2019). Describing her reaction to having to film these scenes in front of an entire set, she offered, "'I have fights on set before when I am like, "No, sheet stays up." And they are like, 'You don't want to disappoint your *Game of Thrones* fans.' And I'm like, "F--- you"'" (Fry Schultz 2019). In her view, if the show was being filmed today, after the #MeToo movement, it would be different in terms of these expectations. Like Selma Hayek, who also spoke up about the way that Harvey Weinstein forced her to do a nude and sex scene with another woman, in order to allow her to continue to make the film *Frida* (2002), other actresses have now come forward to also object to doing this kind of scene. Though Clarke and other women are now more comfortable speaking up, there is still the problem that if an actress is relatively unknown in the industry, they are going to feel pressured to do things they aren't comfortable with, because they want to keep the job. Things won't change, in other words, unless film producers stop asking younger female actresses to do nude scenes, which is unlikely in a media environment where there is still pressure to show these kinds of scenes.

In some ways, the question of nudity also speaks to the larger tensions revealed in the responses by Hollywood to the #MeToo movement. On the one hand, there seems to be a willingness to create storylines that deal squarely with the ramifications of sexual harassment on their characters. On the other hand, there is still a lag in terms of changing the larger culture in more substantive and comprehensive ways behind the scenes where these more progressive storylines are constructed. The example of the television series *The Affair* (Showtime, 2014–2019) is instructive in this regard, both for the frankness with which it dealt with sexual harassment as part of its storyline and how this contrasted with its onset actions with its actors. In Season 5, Episode 8 (original air date, October 13, 2019), the #MeToo scandal forms the basis for bringing back together the married couple Noah Solloway (Dominic West) and Helen (Maura Tierney), who had been ripped apart by Noah's affair from Season 1. Noah is a writer who has been accused by several women of sexual misconduct. The series was premised on the basis of different perceptions about his character, one that portrayed him as volatile but not aggressive. In the face of these accusations, he said, "I work, I write, I see my kids, and I try to be a good person.... You can't just invent reality. Certain things happen, other things don't" (Collins 2019).

As the episode unfolds, however, his relationships with other women, which were sexual and emotionally charged, are questioned as to whether he was predatory. At the same time, his innocence is also questioned, because it shows him acting angry and aggressive and denying that he ever engaged in inappropriate sexual conduct. This was the case, even as he denounces the former student who accused him of just trying to get publicity or going after the ex-publicist at an awards ceremony to confront her about her accusations against him. While the episode portrays Noah as reaching some kind of peace with his ex-wife Helen, it also re-frames his behavior which had been portrayed in the earlier seasons as more equivocal in terms of being sexually aggressive, and now clearly wrong in the era of #MeToo. Some critics found this change in the show's portrayal of Noah as capitalizing on the #MeToo movement and somehow betraying their earlier, more ambiguous attitude toward the character (Collins 2019).[5]

Behind the scenes of the series, other issues that were considered inappropriate in light of the #MeToo movement caused one of the main actresses on the show, Ruth Wilson, to leave the series. Wilson, who earned a Golden Globe for her role as Allison, the woman with whom Noah had an affair, left the show without an explanation in the summer of 2018 (Sandberg and Masters 2019). At the time, she told reporters that she wasn't really allowed to "talk about it," but that they should contact the showrunner Sarah Treem, because "there is a much bigger story" (Sandberg and Masters 2019). *The Hollywood Reporter* followed up with the story, interviewed several people and reported that Wilson wasn't allowed to speak about her departure because she had signed a non-disclosure agreement (Sandberg and Masters 2019). It was subsequently learned that the reason Wilson left the show was because she felt there was a "hostile work environment" and that she believed her character was required to have a great deal of nudity that wasn't warranted by the storylines. While CBS, the parent company of Showtime, had an investigation in 2017 of these charges of a hostile work environment, nothing was subsequently done.

Wilson knew that the show, an adult drama about an affair, would have some nudity in it, but she felt the amount and kind of nude scenes she was required to do often had no rationale other than to titillate the audiences. Wilson had originally signed a nudity waiver when she auditioned for the pilot, but actors and actresses must still provide "meaningful consent" during production of the show. Though Wilson was a vocal critic of the amount and kind of nude scenes she was required to do, Treem still made her perform them, and others who worked on the show believed that, as showrunner, Treem would pressure Wilson and other actors to do the scenes, with one observer on the set being quoted as saying, "Over and over again, I witnessed Sarah Treem try to cajole actors to get naked even if they were uncomfortable or not contractually obligated to," and pressuring them, concluding that "the environment was very toxic" (Sandberg and Masters 2019). Treem, for her part, denied the charges and felt that she had made accommodations for Wilson and that she prided herself on creating strong roles for women. However, other people revealed that there would sometimes be people on the set who didn't need to be there and that they didn't employ an intimacy coordinator until the last season of the show, which would have protected the actors from doing scenes with which they were uncomfortable (Sandberg and Masters 2019).

Ultimately, Wilson left the show after an incident was revealed where Jeffrey Reiner, an executive producer and director of *The Affair*, shared personal information about the nude scenes on the show with *Girls* (HBO, 2012–2017) creator Lena Dunham in September 2016 while they were both on a location shoot for their respective shows. Jenni Konner, a co-creator of *Girls*, recounted a conversation between Dunham and Reiner where Reiner asked Dunham if she would have dinner with Wilson in order to persuade her to "show her tits, or at least some vag" (Sandberg and Masters 2019). After Konner posted this account in the public forum *Lenny Letter*, Reiner ended up meeting with HR, but nothing was done at the time. After the letter was posted, Treem sent an email to the crew of *The Affair* that there was a "zero tolerance policy on sexual harassment and assault," and that while "[t]his is a sexy industry and we are creating a show with a lot of sexual content ... we want to keep that sexy, sexy stuff onscreen. Offscreen, we want to make sure you feel safe and protected while you're working with us" (Sandberg and Masters 2019). At the same time, however, nothing was done to address the comments made by Reiner, even though both Wilson and Tierney were uncomfortable working with Reiner afterward. And, even before the #MeToo movement, in February 2017, Wilson filed a complaint with Showtime that accused them of tolerating a hostile work environment. This is when CBS started the earlier-mentioned internal investigation, but Showtime didn't say how it responded to the investigation or whether anything was done in the wake of it.[6] The way in which the showrunner, the parent company and the cable channel dealt with the issues around a hostile workplace environment reveals the challenges, more generally, faced by those who work in the entertainment industry. Even as television series and films try to tell the story of the women who have been sexually mistreated, the behind the scenes workplace culture still has a long way to go to address the systemic forms of sexism and harassment that co-exist alongside the progressive nature of many of the stories now being shown.

Impact on Film and Television Distribution

Another way to gauge the changes in the entertainment industry in the wake of the #MeToo movement relates to where films are being distributed. Several directors accused of sexual harassment found that, while their films were not being released in the United States, they were still being released overseas. In the case of Woody Allen, his *A Rainy Day in New York* (Amazon Prime Video, 2019) was not able to find a U.S. distributor. In his 53 years of directing films, it was the first time one of his films was not being screened in the United States. Amazon Studios, its U.S. distributor, canceled distribution of the film, and Allen sued the company for $68 million. In its stead, Allen's film ended up being shown in Poland and Lithuania (Schwartzel and Watson 2019) and was released in several other European countries. In addition, Roman Polanski was also unable to find a U.S. distributor for his *An Officer and a Spy* (2019), but was allowed to compete at the Venice Film Festival.

A third director, Nate Parker, who earlier received a strong critical reception at the Sundance Festival for his movie *The Birth of a Nation* in 2016, was similarly denied an opening for his newest film, *American Skin* (2019), in the United States. In 2016, before news came out that he had been charged in a rape allegation, he was able to sell *Birth of a Nation* to Fox Searchlight after its showing at the Sundance Festival for $17.5 million, a huge amount of money for an independent film. Once the earlier rape allegation emerged, however, the film didn't do well at the U.S. box office. *American Skin*, about a Marine veteran who tries to confront the man acquitted in the wake of his son being shot, won the Sconfini Section Prize for Best Film in the 2019 Venice Film Festival. In December 2020, Vertical Entertainment acquired the U.S. distribution rights, and it was released on January 15, 2021, for rent or purchase on a variety of different on-demand platorms, including iTunes, Google Play and FandangoNow. As is the case with Polanski, Parker, too, was able to show his film at the 2019 Venice Film Festival, where they both tried to sell distribution rights to their films. For Polanski and Parker, both of their films deal with people who have been falsely accused or with a miscarriage of justice. In terms of the actual storylines, there are also parallels with Matthew Weiner, who made *Mad Men* and *The Romanoffs*. In the works of Polanski, Parker and Weiner, the themes of false accusations are part of the fabric, and the fact that they are not able to have their films distributed widely in the United States contributes to their sense of receiving unfair treatment in the wake of the #MeToo movement. At the same time, all three of these directors have been able to have their works shown in either domestic (in the case of Weiner) or international markets, as well as film festivals. In these ways, the inconsistency of the responses to these filmmakers mirrors the larger unevenness with which the industry has responded to the #MeToo movement.

#MeToo and the Question of Social Change

One question which emerges is whether the movement is better understood as a social movement and, if so, how that illuminates our understanding of its impact on the film and television industry. It could be argued that Burke's original "Me Too" hashtag that was adopted by Milano was some kind of a social movement, understood by theorists such as Della Porta and Diani (2006) as a process that includes being engaged in a conflictual relationship with clearly identified actors, as well as being connected to informal networks and sharing a clear collective identity. In this understanding, #MeToo doesn't really fit into a classic definition of a social movement. At the same time, other scholars such as Meredith Clark (2016) do make a case for online communities that are formed as a kind of activism, or "hashtag" feminism that can help to create new social paradigms by developing responses to social circumstances they want to see changed. Thus, while the original intent of the #MeToo campaign initiated by Milano was to bring attention to the problem of sexual harassment and assault, it was arguably incredibly effective as a kind of discursive activism that brought to light the magnitude of the problem faced by women in a multitude of industries. As we have seen, though, it is still an open question as to what the ultimate effect will be of the #MeToo movement in changing the hiring practices and stories the entertainment industry will engage in moving forward.

One additional point relating to trying to understand the impact of the #MeToo movement on the entertainment industry is to think about the role of feminism and feminist rhetorics employed to make sense of these narratives. While there is a long history of feminist media analysis, the #MeToo movement has not only served as a source of feminist activism, but also as a way to understand how women's issues are being translated into larger cultural discourses found in film and television. Moeggenberg et al. (2018), for example, have used this understanding of feminist narratology to look at how the recent televisual adaptation of Margaret Atwood's *The Handmaid's Tale* (Hulu, 2017) drew on contemporary discussions about power, consent and the role of ritualized rape. As they offered, "drawing on the dialectic between the #MeToo movement and the series, we discuss the ways in which contemporary protest and feminist activism has integrated key terms and concepts that draw our attention to power, consent and the body" (Moeggenberg et al. 2018). They noted that the #MeToo movement has served as a source of activism both online and in civil society, and that television has also been able to draw on this to question how sexual violence against women has been normalized. *The Handmaid's Tale*, in their view, by showcasing how women have struggled with power and consent, was able to serve as a powerful representation of rape culture in its dystopian vision of a future society. Even though it was written in 1985, the 2017 adaptation of the story by Hulu helped to make it an important vehicle for articulating the contemporary concerns of its audience who were themselves confronting the problem of sexual harassment and assault.

More generally, these examples demonstrate that the #MeToo movement, even as we are still in the middle of it, has been able to provide a language for understanding sexual harassment, and the purpose of this study has been to explore how this has translated into contemporary film and television narratives as well as how it has affected the culture of the entertainment industry responsible for creating these narratives.

Telling and Hearing Stories in a New Light

It is perhaps in the way viewers ultimately understand the stories they are viewing that the influence of #MeToo will have the most lasting impact. Before the #MeToo era, romance and consent were portrayed in film and television in a way that would now be considered problematic. Writers like Julie Beck (2018) have observed that romance usually involves some kind of pursuit, and that the fictional stories we grew up with equated men's desire with a kind of "intrusive attention" the woman was supposed to view as proof of the man's love for her. These images pervaded film and television, and Beck herself recalled growing up "watching movies in which women found it flattering when their pursuers showed up uninvited to hold a boombox under their window, or broke into their bedrooms to watch them sleep, or confessed their feelings via posterboard while their love interest's husband sat in the next room" (Beck 2018). In the #MeToo era, however, it is not just the predatory behavior of Hollywood players behind the scene which is now being re-evaluated, but the ways in which sexual harassment and assault were mythologized in earlier eras.

With a new lens for thinking about these earlier portrayals, viewers can retroactively read them as helping to perpetuate a culture that normalized abusive behaviors. As Beck concluded, "The narratives of a culture help to set its norms. Research has already found that romantic comedies can normalize stalking behavior. It's not difficult, then, to imagine that toxic love stories can also normalize coercion. That they can make people—women, especially—question when and whether their boundaries have really been violated, when they should be flattered and when they should be afraid" (Beck 2018). There are countless examples of how characters who had previously stalked or even raped their love objects get rehabilitated in the context of the storylines, especially when they unfold over several seasons. But in light of what we now understand about the dynamics of sexual violence in the wake of the #MeToo movement, these characters and storylines are no longer taken at face value. And reciprocally, the storylines, which used to normalize behavior around male pursuit and harassment, have been shifting, as witnessed by the raft of new storylines and characters who are portrayed in a negative light when they behave aggressively toward the people they are pursuing.

And, for the stories that are now being told, it is clear that the #MeToo era has brought forth new storylines and characters that can now be read in a new light. From television skits on *Saturday Night Live*'s season premiere in 2018, which had a spoof on the Supreme Court nominee Brett Kavanaugh, to comedies like Netflix's *Unbreakable Kimmy Schmidt*, which had her character, a victim of sexual assault who become an unintentional sexual harasser in the workplace, both storylines and characters have drawn on the themes of #MeToo to drive their narratives. These stories, in turn, are a reflection of the ways the larger culture is finally turning its attention to the problem of sexual harassment, or as Marsha Barber has noted, "The fact that society is now addressing sexual misconduct, harassment and abuse head-on gives pop culture license to do the same" (Dziemianowicz 2018). And, as we have seen, comedies have played an important role in this reckoning in popular culture, for a variety of reasons, including its capacity to both mock and reveal larger cultural anxieties around social change, including relations between the sexes. David Hinckley has found that comedies have been able to address these issues in a particularly thoughtful manner, from *Unbreakable Kimmy Schmidt* to *Great News*, where the main character harasses her staff so she can get a high pay package to leave her job, as high profile male harassers before her were able to do. Hinckley's reading is that this is indicative of the way comedies can speak to more serious issues around harassment, or as he offered, "Those storylines could easily have felt like uneasy cheap laughs, but they didn't because each also made a serious point: that we're more aware of inappropriate behavior these days, and that guys who behave inappropriately even now can escape with no consequences or even a seeming reward" (Dziemianowicz 2018).

At the same time, dramas also revealed the tensions that were raised in the wake of #MeToo allegations and provided fuel for plotlines that centered on the effects of harassment on the characters who either committed the acts or were victims of them. In the genre of horror films as well, storylines were created with backstories of sexual trauma and abuse as a way to gain entry into the motivations of predators who were now often women who became vengeful in the face of the abuse they experienced earlier in their lives. Reality shows, and in particular, reality shows that centered on dating or being in stressful physical situations, were also revelatory in terms of their oftentimes unintended encounters with sexual misconduct that took place on their shows, and the responses revealed a complicated set of half-measures that both tried to reign in the excesses of these acts and at the same time to capitalize on them for ratings. On the international level, different countries' film and entertainment industries encountered the rise of voices protesting sexual mistreatment with a variety of responses, which oftentimes was reflective of the particular country's cultural attitudes toward women. In France, for example, there was a virtual explosion on the internet and in conversations in families about whether #MeToo had gone too far, and there has been an ongoing debate about the excesses of the movement that still has not been resolved. At the same time, however, the rise in women coming forward to file sexual harassment complaints does indicate that the links between cultural change and changes in the film and entertainment industry often go hand in hand.

In the end, it is this insight that the pace of cultural change is both mirrored and reflected in the entertainment industries of Hollywood and beyond that leaves the question ultimately open whether #MeToo has permanently changed the landscape of Hollywood and the international film and television industries. The structural problems of Hollywood's lack of women being represented in positions of power and in creative roles revealed in this study indicates that it will take many years to reverse this course. Some of these structural problems are being addressed with new guidelines around codes of conduct on the set and the hiring of intimacy coordinators. In addition, there are legal changes restricting the use of non-disclosure agreements, as well as corporate changes and firings of individuals who engaged in egregious forms of harassment. There has also been social activism and the creation of the Commission for Eliminating Sexual Harassment and Advancing

Equality. And finally, in terms of representation, we are beginning to see more women hired for roles as showrunner, directors, writers and other creative roles, as well as more stories being told from a woman's point of view.

While the structural changes may be slow in coming and may take years to be felt, in the meantime, we are still dealing with the reality of what these allegations have meant for the larger society. We are still confronting the structural basis of the forms of inequality which gave rise to unequal power relations and which allowed sexual harassment to go on for decades unabated. Jia Tolentino has made the larger point that the #MeToo movement has shown that the power of the story of victimhood by those women who have come forward is being challenged by a counter-narrative that the individuals who have been accused are themselves victimized. As she wrote, "The #MeToo movement has demonstrated the principled urgency of a persuasive story of victimhood; the accused now crave that righteousness, too. They want to possess the moral power that their victims are wielding over them, along with the structural power they have long had and still possess. It's the original demand, replicating itself fiendishly: whatever is yours must belong to me" (Tolentino 2020). For Tolentino, this ability to shapeshift the truth is one of the ways the #MeToo movement has been undermined. Another way is to pretend that the #MeToo movement has "already won," and that "the centuries of male domination have been swiftly reversed to make women the rulers of the land" (Tolentino 2020).

It is this insight, perhaps, that the impact of the #MeToo movement on film and television will be undermined by declaring that it has already won that I think is the most helpful way of moving forward, both in terms of the industry itself as well as the larger culture. For, no matter how powerful the claims of the women who have been mistreated are, it is a long road to making permanent changes in the entertainment industries. But as we have seen, for all of the back steps and countermeasures, there have been some changes made, and more stories are finally being told. There has been a virtual deluge of plotlines centered on #MeToo themes, and these plots reflect stories that are happening in the larger culture. As Emily Nussbaum has observed, television shows and those who create them are never unaware of their audiences, because if they are, "they don't get renewed" (Nussbaum 2019). This can account for what she describes as their strength but also their limitation.

As the creators of these shows try to engage with the subject of #MeToo, both because it gives currency to their shows as well as because they want to move the needle on the subject, it can be said that they are still in the early stages of representing these cultural shifts.

For Nussbaum and others, this sense that we are still in the early stages means that there will be things missing and that there are stories that have yet to be told from more genres, including science fiction. The sense that "the lens can keep widening" is a hopeful note to end on, as well as the idea that the best film and television can reveal the ambiguity of the times we are living in and can provide "a minor-key note inside a culture-wide chord" (Nussbaum 2019). It is these notes that we can observe with more frequency as the culture itself widens on a world that will hopefully continue to fight for the struggles represented by the women who were brave enough to come forward to tell their truth.

Chapter Notes

Introduction

1. In 2006, Burke founded a nonprofit organization called Just Be Inc., with the phrase "MeToo," which she said came to her in 1997 after she heard a 13-year-old speak about an experience of sexual abuse. Burke told the *New York Times* that she didn't have a way to help the young woman at that moment, and that she couldn't say the words "me too" at the time (Nicolaou & Smith 2019).

2. Despite this, Franco went on to win a Golden Globe that year for his role in *The Disaster Artist* though he ended up being snubbed for an Oscar nomination.

3. Another theorist, Orgad (2014), reflects more deeply on this dynamic, looking at how media is both a reflection of societal understandings as well as helping to construct those very relationships.

Chapter 1

1. As Tarana Burke, the woman who originally used the hashtag #MeToo, noted, "The celebrities who popularized the hashtag didn't take a moment to see if their work was already being done, but they also were trying to make a larger point" (Hutchinson 2018).

2. Looking at data from the U.S. Equal Employment Opportunity Commission, for example, the government agency responsible for investigating complaints of workplace harassment and discrimination, found that they had over 7,500 harassment complaints filed between October 2017 and September 2018, which was a 12 percent increase from the number filed the previous year (Chiwaya 2018). This was the case despite the fact that overall complaints had dropped. Victoria Lipnic, acting chair of the EEOC, noted that visits to the EEOC's sexual harassment page had doubled from the previous year as well, stating "the impact of the #MeToo movement is undeniable" (Chiwaya 2018).

3. Because of the surge of women who came forward to tell their stories and file harassment claims, there were over 41 harassment lawsuits that the EEOC eventually filed in 2018, which doubled the number filed the year before. The companies the EEOC filed claims against included both large corporations as well as smaller companies. These companies, from United Airlines to small stores like a Dollar General in Maryland, had not previously admitted any wrongdoing. Looking at these cases, it is clear that while the type of case is not new, what is new and what is based on the impact of the #MeToo movement, is that the number of cases increased exponentially. In addition to these cases, victim resource hotlines also found that there was a spike in calls to them. Safe Horizon, for example, a victim-assistance nonprofit with a domestic violence, rape and sexual assault hotline, found that they had a 52 percent increase in the number of calls they received from the year before and that on September 28, 2018, they received a 500 percent spike in the number of calls received. This was the day after Christine Blasey Ford testified to the Senate committee detailing her allegation that Brett Kavanaugh had sexually assaulted her

when they were teenagers (cited in Chiyawa 2018). While Ford's testimony was not directly part of the #MeToo movement, the fact that it occurred in the year after this movement began speaks to the kind of national conversation invoked in the wake of this movement.

Chapter 2

1. The title of the show itself refers to the fact that after the rape, Dr. Melfi sees a picture of the rapist on the wall of a local sub shop where she was ordering her lunch and it said: "Employee of the Month" underneath his picture.

2. Apparently, to many readers of the comic, this scene was both disturbing and felt like it was not part of the rest of the story (Outlaw 2019).

3. This trope of dead females, typically young and white, is so common that most viewers know there will invariably be an opening scene with a dead female, as seen on such varied shows as *Breaking Bad*, *Dexter* and Netflix's *The Punisher*.

Chapter 3

1. Merkin also provocatively asks at one point in the article, whatever happened to flirting? (Merkin 2018).

2. These comments echo some of the ideas of his debut novel, *Bob Honey Who Just Do Stuff* (2018), where he referred to the #MeToo movement as "an infantilizing term of the day," and where he asked whether the movement was a "toddler's crusade," which ends up "reducing rape, slut-shaming and suffrage to reckless child's play" (Sharf 2018b).

3. As he observed, "There are some famous people being suddenly accused of touching some girl's knee or something, and then suddenly they've been dropped from their program" (cited in Nordine 2018a).

4. Affleck continued that he was scared because of how he was raised and that the values he was raised by are the same as the values at the heart of the #MeToo movement, as for example when his mother wouldn't let him watch an earlier

television show like *The Dukes of Hazzard* because she thought it was sexist (Sharf 2019b).

5. As he offered on twitter, "tting right there! I'm incredibly embarrassed and deeply sorry to have done that to Jessica. This is a big learning moment for me. I shouldn't have tried so hard to mansplain, or fix a fight, or make everything okay. I should've focused more on what the most important..." (Deb 2018b).

6. As one viewer wrote on Twitter, "It wasn't a fight, was it? It was a man screaming at a coworker so viciously that years later, she is still shaken & tearful. Do you notice how fragile/volatile Tambor is, that you felt compelled to soothe/ puff up/calm/praise him? Keep self assessing, bc you don't get it yet" (Deb 2018b).

7. After the controversy over the interview with the cast of *Arrested Development*, Netflix did cancel a cast promotional tour in England that was scheduled for the following week. But Netflix continued to show their faith in the production and it stayed on the air.

8. The excuse, however, comes through clearly when he continued with "what was previously accepted is now untenable to anyone of a certain consciousness" (2017b). In this way, in the same interview, he acknowledges his own complicity in tolerating these behaviors, and also identifies it as part of an earlier time when cultural standards were different. Perhaps most tellingly, when Kantor asked Tarantino how he thought the Weinstein allegations will affect how the public perceives Tarantino's own work, he told her "I don't know...I hope it doesn't" (Kantor 2017b).

9. As Cassie da Costa wrote for *The Daily Beast*, "But the more urgent, and meaningful, conversation is about those who have not yet succeeded in the industry, from up-and-coming comedians to struggling actors to the staff who serve at the bars, clubs, and festivals where these men tend to trickle back in, and how more space might be made for their voices, dissent, ideas, disavowals. If action around #MeToo within the entertainment industry is only interested in the performers who already have leverage— SAG and WGA members, TV regulars, award-winners, and movie stars—then

the abuses will simply settle more deeply into the more vulnerable places they fester: the clubs, showcases, and modest backstage corridors defined by the mix of wild ambition and economic precarity of the performers who make their way through them" (DaCosta 2019).

10. Other women have spoken out as well about the hire, including Mireille Soria, the head of Paramount's animation division. Because Skydance has a partnership with Paramount, this was an important move because Soria, a respected executive and a long-time associate of animation pioneer Jeffrey Katzenberg, would no longer do business with Skydance. And the fact that Thompson had the courage to speak up and leave a film project meant that Skydance might have trouble in the future of bringing other talent and executives who would have to work with Lasseter on an ongoing basis.

Chapter 4

1. Or as Shukert concluded, "you have this side of you that wants to be outraged and complain, and then you have the other side of you that wants to be tough and wants to be cool.... I deeply felt both of their points of view, and we felt like both sides had validity. That Debbie has done what she had to do ... and Ruth did the right thing, and now, unfairly, they all have to suffer the consequences. It's the conversation between these women that is always important on our show.... The female gaze, even when it applies to other females" (cited in Miller 2018).

2. Though he was never formally charged with anything, a former junior colleague of Matt Lauer's did, however, accuse him of raping her in his hotel room in 2014, at the Winter Olympics in Sochi, Russia. She first discussed this publicly in an interview with a journalist, Ronan Farrow, and it was then published in his book, *Catch and Kill: Lies, Spies, and a Conspiracy to Protect Predators* (2019). The woman, Brooke Nevils, told Farrow that Lauer had forced her to engage in anal sex without her consent. This complaint was what ultimately led to Lauer being fired from *The Today Show* in 2017, but it wasn't until October of 2019 that the specific allegations came to light. When advanced copies of Farrow's book came out, NBC News made a statement, saying Lauer's conduct was "appalling, horrific and reprehensible" (cited in Folkenflick Dwyer 2019). Lauer responded to the charges that came to light in the book by saying that the relationship had been consensual and that he never raped her.

3. One television critic, Meredith Blake, has referred to this defense as the "at least I'm not Harvey Weinstein," excuse, and cites other entertainers such as Louis C.K., who had admitted to masturbating in front of several women without their consent, but who believe that they never did what Weinstein did (Blake 2019).

Chapter 5

1. As he offered, "'The challenges and the difficulties and the frustrations—we wanted to embrace it and we wanted it to carry forward,' Shore says. 'We didn't want it to be something that happens and then it's just forgotten about. We wanted it to play out over a few episodes. We didn't want it to be what the show is completely all about, but we wanted to deal with it honestly, and that means for more than just a moment. This is something that carries on and that people live with'" (cited in Friedlander 2018).

2. Noting the frequency of the attacks the character Meredith suffers, Horan wrote, "Whatever your thoughts are on the current state of *Grey's Anatomy* (I actually think the show has turned a corner this season, with enough fresh faces mixed in with veterans to start returning to its roots), it has created some thrilling moments of TV, and some of its most effective episodes have been rooted in danger and violence. But Meredith's upcoming beating seems, at this point, redundant. She's already been beaten. If she visited another hospital, even for a checkup, they wouldn't get through her medical history before locking her up in a safe house for her own protection. So much bloody bad luck shouldered by one character stretches the limits of the audience's suspension of disbelief—it's getting

old, and it's getting draining. In a still from the upcoming attack video, Meredith, bloodied on the floor, looks more tired and annoyed than hurt. We feel you, Mer" (Horan 2016).

3. As one writer for Refinery29.com noted: "Not everyone thought justice was served, though, pointing out that Catherine was ultimately left off the hook for her role in protecting her father's bad behavior. At first, Catherine's lawyer even suggested not dissolving the foundation and instead attacking the women who had come out against him using the excuse 'someone is going to take the fall.' The lawyer just hoped that someone wouldn't be Harper Avery" (Carlin 2018).

4. As one online user of Reddit commented, Catherine "should have to pay some price because she helped cover up what HA was doing." Adding, "Not only do those women get to suffer the trauma of sexual harassment/violence and lose the opportunity to win the highest honorable award in the medical field, but now they have to sit back and watch while the person who helped cover up their attacks and shut them up, rises to an even higher top?!" (Carlin 2018).

5. Writing about the difference between *The Good Fight* and *The Good Wife*, television critic Shannon Miller noted "Christine Baranski ostensibly stars, carrying the baton from 'The Good Wife' like the legend she is. But she's only one leg of the tripod that supports 'The Good Fight,' following the episode's inciting incident: A Bernie Madoff–esque Ponzi scheme that threatens to ruin the lives of both Baranski's Diane Lockhart and newly minted lawyer Maya (Rose Leslie). There's also Cush Jumbo, a wonderful find from Britain who appeared in 'Good Wife' Season 7, and returns with a really fabulous haircut and a wonderfully no-bullshit attitude. And fuck yes: It's delightful to watch a show jam-packed with these strong women" (Miller 2017).

6. Michelle King described the impetus for doing this kind of show thus: "One can certainly do episodes of television about someone that breaks into a home and rapes a woman, but I don't think there is any question about knowing that's wrong.... To be in areas where folks aren't recognizing that their behavior is wrong

is far more interesting to us" (cited in McHenry 2019)

7. One of the writers for the show, Joey Heartsone, had also gotten his start on a reality show, *Project Runway*, and was also able to bring a "layer of reality" to the episode (cited in McHenry 2018).

8. They referenced an event that happened in 1921, when a young American actress named Virginia Rappe was attacked by an actor, silent film star Roscoe "Fatty" Arbuckle. Arbuckle was charged with manslaughter when Rappe died four days after the attack. Arbuckle was acquitted in the third trial, after the first two cases had a hung jury.

9. As Jelani Cobb, a writer for *The New Yorker*, described the documentary: "An exposé typically indicts the character of its subject; 'Surviving R. Kelly' indicts a public that knew of his character and did nothing about it, a public that constructed an elaborate architecture of denial and has chosen to live in it" (cited in *The New Yorker* Interview 2019).

10. Roach describes being able to generate some of the ways Ailes harassed potential female anchors, and referred to the same way that he would have them show him their behind, by asking them to "stand up and give me a twirl" (cited in Walsh 2019).

11. Citing her culpability, Mitch asks her, "'Are you going to keep pretending you didn't know what was going on?' he asks. 'Are you actually going to look me in the eye and say you didn't participate? ... You didn't roll your eyes at these women? You didn't make jokes at their expense? You didn't mock their sometimes desperate behavior when I moved on? I may have f-cked them, but you were very cruel. And words matter'" (cited in Dockterman 2019).

Chapter 6

1. They then went on to note: "A formal warning was also given to the male castaway in question. On *Survivor*, producers provide the castaways a wide berth to play the game. At the same time, all castaways are monitored and supervised at all times. They have full access to producers and doctors, and the production will

intervene in situations where warranted" (cited in Bradley 2019 (b)).

2. Or as Poniewozick offered, "Watching 'Survivor' bungle Kim's complaints, well into the #MeToo era, was like watching a recurring nightmare: A woman is touched inappropriately, she speaks up about it, her concerns are minimized or paid lip service.... Oh, but she'd have been treated better if only there were proof, right? Ha ha, guess again! Even when there is video documentation—even on a show whose premise is constant surveillance—the behavior still continue and the business that she complains to still does next to nothing. What's more, she's the one who suffers for speaking up" (Poniewozick 2019).

3. Poniewozick also noted the tone deafness of CBS in other examples, as when he observed that "CBS has acted deaf, dumb and blind in handling cases on individual shows, keeping on Michael Weatherly, the star of 'Bull,' even after it paid a $9.5 million settlement to the actress Eliza Dushku, who said she was axed from the show after complaining about Weatherly's inappropriate comments. CBS fired a former showrunner of 'N.C.I.S.: New Orleans,' but only after a long history of inappropriate-behavior allegations that the network has been accused of creating a culture in which harassers and misogynists thrived" (Poniewozick 2019)

Chapter 7

1. Heidi Linden, a Finnish film director, has collected the stories of some 40 women, and in these stories, the same 5–10 men are routinely mentioned. At the same time, the overall response to #MeToo has not resulted in any substantial changes in the industry (Hild 2018).

2. The summit ended up aligning with the group called Screensafe, which had been set up in 2015 to make sure that the New Zealand screen industry had some health and safety standards in place.

3. Though there is one notable exception to indicate some response to victims. In December 2018, for example, producer-director Bey Logan, who ran the Weinstein Co.'s Asia office from 2005 to 2009 was accused of sexual misconduct by multiple women. In Hong Kong, this signaled that the #MeToo movement had gained some momentum, even as there are still cultural barriers, including victim shaming and lack of legal support.

4. He explained what he hoped to achieve in his three short films, noting "So we have male perspective, female perspective and a kind of K-pop theme.... I say 'loosely' [built around a #MeToo theme] because I'm not interested in the propaganda approach. We have that in multi-forms everyday in newspapers and on TV. It's really about examining the issues on a more intimate social level, through a couple of macro moments" (Scott 2019).

Conclusion

1. Weinstein has been accused by them of a range of behaviors, from sexual harassment to outright rape. Although the punishment to Weinstein as well as the fact that he will not be personally liable is deeply troubling to many of them, they were risking not getting anything at all if they had not settled or held out for different terms.

2. Kantor also noted that a lot of the money will go directly to the lawyers for the plaintiffs. Part of the problem is that Weinstein's company filed for bankruptcy, and the creditors for the company also were due a portion of the money from the company. The money the victims will be paid will not come from Weinstein, but from The Weinstein Company's insurance policies. And, even if people are outraged that an insurance company will pay for sexual harassment claims, this is how the insurance policy works.

3. Kathryn Bigelow was the first woman to win that category in 2010 for *The Hurt Locker.*

4. One of the purposes of the survey is to allow for anonymity so that people can respond truthfully to their experiences in the industry. The survey will ask whether they have experienced inappropriate conduct, whether they were threatened if they didn't comply with inappropriate behavior and whether their employers

were clear about what was considered appropriate and inappropriate behavior. The survey will be conducted by a collaboration of nonprofits called The Ethics and Compliance Initiative, and the report will be released in 2020 (Shaw 2019).

5. Sean Collins, for example, writing for the *New York Times*, had this analysis of the episode's perspective, which was one of the last episodes of the series' five season run: "But I can't shake the feeling that the show is backfilling a #MeToo payload into a space it was content to leave undeveloped until just now. While individual incidents involving Alison and other women drove the occasional episode or arc, a coherent Noah-as-obvious-serial-predator narrative is new. Considering how many different vantage points we've had into Noah's life—his own, his ex-wives', his girlfriends', his daughter's, and even that of a guy who once pointed a gun at him in anger—to have these accusations emerge now feels like a narrative cheat" (Collins 2019).

6. During the time of the investigation, Reiner was not allowed to work with Wilson on any episodes but could work on other episodes, and he decided to leave the show after the third season. While he left *The Affair*, he was still considered for other Showtime dramas, including *I'm Dying Up Here*, and he directed an episode of *Shameless*, also put out by Showtime. The fact that he was still considered for and worked on other Showtime series, created a sense for the actors that the allegations against him weren't being taken seriously. One of the effects of this incident, however, was that Wilson was allowed to leave the show, and was able to have her perspective taken into account for what would happen to the character, though it was limited. She was originally told that the character would be subject to a violent sexual assault and murder, and in the end, the character was murdered, but not sexually assaulted. For the actress, the ending she had wanted for her character was to "walk into the sunset with her son and with no man" (cited in Sandberg Masters 2019).

Bibliography

Abramson, A. (2018). "Brett Kavanaugh Confirmed to Supreme Court After Fight That Divided America." *Time Magazine,* October 7. Retrieved from https://time.com/5417538/bett-kavanaugh-confirmed-senate-supreme-court/.

Anderman, N. (2017). "#MeToo Shakes Up Israeli TV and Film Industry as Victims of Harassment Speak Out." *Haaretz,* October 18. Retrieved from https://www.haaretz.com/israel-news/.premium-harassment-of-women-prevalent-in-local-film-and-tv-industry-1.5458598.

Anderson, J. (2019). "The #MeToo Movement Gets a Movie About Everyday Harassment with Israel's 'Working Woman.'" *Los Angeles Times,* April 12. Retrieved from https://www.latimes.com/entertainment/movies/la-et-mn-working-woman-20190412-story.html.

Angelo, M. (2018). "Kimmy Schmidt Is TV's Most Unlikely #MeToo Villian." *Glamour,* June 7. Retrieved from https://www.glamour.com/story/kimmy-schmidt-metoo018June.

Avrich, B. (2018). "Hollywood's Infuriating Journey from #MeToo to #TooSoon: John Lasseter's Possible Return to Disney Reveals a Sad Forgive-and-Forget Mentality That Insults the Women Who Have Suffered, Writes the Director of a New Sexual Harassment Documentary." *The Hollywood Reporter* 424(19): 48.

Barnes, B. (2018). "A Year After #MeToo, Hollywood's Got a Malaise Money Can't Cure." *New York Times,* November 8. Retrieved from https://nyti.ms/2z0BGq9.

Beck, J. (2018). "When Pop Culture Sells Dangerous Myths About Romance." *The Atlantic,* January 17. Retrieved from https://www.theatlantic.com/entertainment/archive/2018/01/when-pop-culture-sells-dangerous-myths-about-romance/549749/.

Bennett, A. (2018). "How #MeToo Changed Some of Your Favorite TV Shows in 2018." *BuzzFeed News,* October 8. Retrieved from https://www.buzzfeednews.com/article/alannabennett/metoo-weinstein-tv.

Bernstein, A. (2018). "Small Screen Survivors: How US TV Is Handling the #MeToo Movement." *The Guardian,* June 25. Retrieved from https://www.theguardian.com/tv-and-radio/2018/jun/25/small-screen-survivors-how-tv-is-handling-the-metoo-movement.

Birnbaum, D. (2018). "'Murphy Brown' Creator on Moonves Allegations: 'We Support the Investigation.'" *Variety*, August 5. Retrieved from https://variety.com/2018/tv/news/murphy-brown-creator-on-moonves-allegations-we-support-the-investigation-1202895550/.

Blake, M. (2019). "How Netflix's 'Unbelievable' Created Its Revolutionary Portrayal of Rape." *Los Angeles Times,* October 3. Retrieved from https://www.latimes.com/entertainment-arts/tv/story/2019–10–03/unbelievable-netflix-rape-representation.

Blake, M. (2019). "MeToo Storyline Drives New Series: *The Morning Show* Dramatizes Complications, Ripple Effects of Movement." *Winnipeg Free Press,* November 9. Retrieved from https://www.winnipegfreepress.

com/arts-and-life/entertainment/TV/metoo-storyline-drives-new-series-564699171.html.

Bradley, L. (2019a). "Changing Hollywood Sex Scenes for Good." *Vanity Fair*, August 16. Retrieved from https://www.vanityfair.com/hollywood/2019/08/intimacy-coordinators-screen-actors-guild?utm_source=nl&utm_brand=vf&utm_mailing=VF_CH_081819&utm_medium=email&bxid=5bea02e13f92a404693d10a4&cndid=8477505&hasha=498fcefa5047809f5dae6bc1d6f91255&hashb=1de06b71a2246c82c0cc8036477336e7cfdd91fa&hashc=c0dc54ef3aff54424ed534a7f38c8d3a64c68c753fecb25cd82d41541b2cd812&esrc=newsletteroverlay&utm_campaign=VF_CH_081819&utm_term=VYF_Cocktail_Hour.

Bradley, L. (2019b). "Will Reality Television Ever Learn How to Handle Misconduct Allegations?" *Vanity Fair*, November 15. Retrieved from https://www.vanityfair.com/hollywood/2019/11/survivor-dan-spilo-kellee-kim?utm_source=nl&utm_brand=vf&utm_mailing=VF_CH_111519&utm_medium=email&bxid=5bea02e13f92a404693d10a4&cndid=8477505&hasha=498fcefa5047809f5dae6bc1d6f91255&hashb=1de06b71a2246c82c0cc8036477336e7cfdd91fa&hashc=0dc54ef3aff54424ed534a7f38c8d3a64c68c753fecb25cd82d41541b2cd812&esrc=newsletteroverlay.

Breeden, A. (2019a). "Sex Crime Reports Are Up in France. Officials See a #MeToo Effect." *New York Times*, February 1. Retrieved from https://www.nytimes.com/2019/02/01/world/europe/france-sex-crimes.html?action=click&module=RelatedLinks&pgtype=Article.

Breeden, A. (2019b). "French Me #MeToo Movement's Founder Loses Defamation." *New York Times*, September 25. Retrieved from https://www.nytimes.com/2019/09/25/world/europe/france-sandra-muller-verdict.html.

Brody, R. (2019). "'Ma' and 'The Perfection,' Reviewed: Two Horror Movies Crassly Exploit #MeToo." *The New Yorker*, May 31. Retrieved from https://www.newyorker.com/culture/the-front-row/ma-and-the-perfection-reviewed-two-horror-movies-crassly-exploit-metoo.

Brzeski, P. (2019). "New Zealand Launches Innovative Program to Prevent Sexual Harassment in Screen Industry." *The Hollywood Reporter*, October 10. Retrieved from https://www.hollywoodreporter.com/news/new-zealand-launches-innovative-program-prevent-sexual-harassment-screen-industry-1246983.

Buckley, C. (2019). "An Idea for Hollywood Has Few Adopters." *New York Times*, June 20. C1(L).

Carlin, S. (2018). "*Grey's Anatomy's* Big #MeToo Moment Has Fans Divided." *Refinery29*, April 27. Retrieved from https://www.refinery29.com/en-us/2018/04/197579/greys-anatomy-metoo-moment-divides-fans.

Castells, M. (2007). "Communication, Power and Counter-power in the Network Society." *International Journal of Communication* 1, no. 1 (2007): 29.

Castillo, M. (2020). "This Is What the Hollywood Backlash to #MeToo Looks Like." *TheLily*, May 10. Retrieved from https://www.thelily.com/this-is-what-the-hollywood-backlash-to-metoo-looks-like/.

Chira, S. (2018). "Numbers Hint at Why #MeToo Took Off: The Sheer Number Who Can Say MeToo." *New York Times*, February 21. Retrieved from https://www.nytimes.com/2018/02/21/upshot/pervasive-sexual-harassment-why-me-too-took-off-poll.html.

Chiwaya, N. (2018). "New Data on #MeToo's First Year Shows UnDeniable Impact." *NBC News*, October 11. Retrieved from https://www.nbcnews.com/news/us-news/new-data-metoos-first-year-shows-undeniable-impact-n918821.

Clark, R. (2016). "'Hope in a Hashtag': The Discursive Activism of #WhyIStayed." *Feminist Media Studies* 16, no. 5, 788–804.

Clark, T. (2019). "How Amazon's New Superhero TV Show, 'The Boys,' Was Shaped by Trump, Me Too, and 'Sweet, Sweet Bezos Money.'" *Business Insider*, July 17. Retrieved from https://www.businessinsider.com/the-boys-showrunner-eric-kripke-interview-trump-me-too-amazon-2019-7.

Clements, S. (2020). "'Portrait of a Lady on Fire' Star Noemie Merlant Discusses

#MeToo and Playing a Woman Outside the Patriarchy." *Exclaim.ca*, February 13. Retrieved from http://exclaim. ca/film/article/portrait_of_a_lady_on_fire_star_no_mie_merlant_discusses_metoo_and_playing_a_woman_outside_the_patriarchy.

Cobb, S., and Horeck, T. (2018). "Post Weinstein: Gendered Power and Harassment in the Media Industries." *Feminist Media Studies* 18, no. 3, 489–491.

Cohen, A. (2019). "Revisiting *The Sopranos* Most Controversial Episode Ever." *Refinery29*, January 10. Retrieved from https://www.refinery29. com/en-us/2019/01/221249/the-sopranos-anniversary-women-abuse-season-3-episode-6.

Collins, L. (2018). "Why Did Catherine Deneuve and Other Prominent French Women Denounce #MeToo?" *The New Yorker*, January 10. Retrieved from https://www.newyorker.com/news/daily-comment/why-did-catherine-deneuve-and-other-prominent-frenchwomen-denounce-metoo.

Collins, S. (2019). "'The Affair' Season 5, Episode 8: He Said, She Said, She Said, She Said." *New York Times*. October 13. Retrieved from https://www.nytimes.com/2019/10/13/arts/television/the-affair-recap.html.

Combemale, L. (2019). "The Creators of Netflix's *Unbelievable* on Their Urgent New Series." *The Credits*, September 13. Retrieved from https://www.motionpictures.org/2019/09/the-creators-of-netflixs-unbelievable-on-their-urgent-new-series/.

Coyle, J. (2018). "#MeToo: Hollywood Is Still Soul Searching a Year After Weinstein." *AP*, October 3. Retrieved from https://apnews.com/20d161014e3c46dfa5dea469b14f34e1.

Crispin, J. (2020). "How Hollywood Handles Women in Peril." *Variety*, March 26. Retrieved from https://variety.com/2020/film/features/how-hollywood-handles-women-in-peril-1203545055/.

da Costa, C. (2019). "Harvey Weinstein's Actors Hour Fiasco and the Illusion of 'Cancel Culture': The Accused Serial Rapist/Disgraced Movie Mogul's Appearance at an Event for Young Actors, and How It Was Handled, Says a Lot About How We Treat Vulnerable Women and Powerful Men." *The Daily Beast*. October 30. Retrieved from https://www.thedailybeast.com/harvey-weinsteins-actors-hour-fiasco-and-the-illusion-of-cancel-culture.

Deb, S. (2018a). "'Arrested Development': We Sat Down with the Cast. It Got Raw." *New York Times*, May 23. Retrieved from https://nyti. ms/2J2kKph.

Deb, S. (2018b). "Jason Bateman Apologizes to Jessica Walter Over Jeffrey Tambor Comments." *New York Times*, May 24. Retrieved from https://nyti. ms/2xfkPRM

Della Porta, D., and Diani, M. (2006). *Social Movements: An Introduction*. Blackwell.

DeRogatis, J. (2019). "R. Kelly and the Damage Done." *The New Yorker*, June 3. Retrieved from https://www. newyorker.com/culture/personal-history/r-kelly-and-the-damage-done.

Dockterman, E. (2019). "*Bombshell* Is the Latest Example of Pop Culture Reckoning with the Complicated Reality of #MeToo at Work." *Time*, December 16. Retrieved from https://time. com/5744161/bombshell-succession-the-morning-show-me-too/.

Dwyer, C. (2020). "Harvey Weinstein Sentenced to 23 Years in Prison for Rape and Sexual Abuse." *NPR*, March 11. Retrieved from https://www.npr. org/2020/03/11/814051801/harvey-weinstein-sentenced-to-23-years-in-prison.

Dziemianowicz, J. (2018). "How #MeToo Has Changed Pop Culture Forever." *MarketWatch*, October 3. Retrieved from https://www.marketwatch. com/story/how-metoo-has-changed-pop-culture-forever-2018–10–03?reflink=MW_GoogleNews.

Edelstein, R. (2019). "'Brockmire' EP Says No Comedy Is Off Limits When Directed at 'Stupid Incompetent People in Power'—AMC Networks Summit." *Dateline*, April 8. Retrieved from https://deadline.com/2019/04/comedy-metoo-era-limits-brockmire-odd-mom-out-producers-amc-networks-summit-1202591641/.

Erbland, K. (2018). "Laura Dern: #MeToo

Has Already Changed the Way Hollywood Treats the Next Generation." *IndieWire*, June 13. Retrieved from https://www.indiewire.com/2018/06/laura-dern-me-too-changed-hollywood-1201973946/.

Farrow, R. (2017). "From Aggressive Overtures to Sexual Assault: Harvey Weinstein's Accusers Tell Their Stories." *The New Yorker*, October 23. Retrieved from https://www.newyorker.com/news/news-desk/from-aggressive-overtures-to-sexual-assault-harvey-weinsteins-accusers-tell-their-stories.

Faughnder, R., and Perman, S. (2020). "Five Things That Have Changed in Hollywood Since the Weinstein Case Broke." *Los Angeles Times*, February 24. Retrieved from https://www.latimes.com/entertainment-arts/business/story/2020–02–24/five-things-that-have-changed-in-hollywood-since-the-weinstein-case-broke.

Fienberg, D. 2019. "'90s Nostalgia Meets the #MeToo Movement: TV's Obsession with the 1990s Has Mostly Manifested Itself in Reboots and Remakes, but Three Docuseries About Sexual Abuse Suggest a Darker, Timelier and More Progressive Prism Through Which to View the Decade." *The Hollywood Reporter*, February 20. p. 211.

Folkenflik, D., and Dwyer, C. (2019). "Matt Lauer Accused of Rape in New Book; Former NBC Former NBC Star Denies 'False Stories.'" *NPR*, October 9. Retrieved from https://www.npr.org/2019/10/09/768527936/matt-lauer-accused-of-rape-in-new-book-former-nbc-star-denies-false-stories.

Freeman, H. (2018a). "What Does Hollywood's Reverence for Child Rapist Roman Polanski Tell Us?" *The Guardian*, January 30. Retrieved from https://www.theguardian.com/film/2018/jan/30/hollywood-reverence-child-rapist-roman-polanski-convicted-40-years-on-run?

Freeman, H. (2018b). "Actors Are Lining Up to Condemn Woody Allen. Why Now?" *The Guardian*, February 3. Retrieved from https://www.theguardian.com/global/2018/feb/03/actors-condemn-woody-allen-hadley-freeman.

Friedlander, W. (2018). "Amplifying the Voice of a Movement." *Variety* 340, no. 9 (June 7, 2018): 8.

Fry Schultz, M. (2019). "Emilia Clarke Reveals Hollywood's Other #MeToo Problem." *The Examiner*, November 20.

Fuchs, C. (2017). *Social Media: A Critical Introduction*. Sage.

Garber, M. (2018). "The Selective Empathy of #MeToo Backlash." *The Atlantic*, February 11. Retrieved from https://www.theatlantic.com/entertainment/archive/2018/02/the-selective-empathy-of-metoo-backlash/553022/.

Garber, M. The Atlantic Culture Desk. (2019). "The 7 Most Defining #MeToo Moments of 2019." *The Atlantic*, December 12. Retrieved from https://www.theatlantic.com/entertainment/archive/2019/12/most-defining-metoo-moments-2019/603490/.

Gilbert, S. (2018a). "The Men of #MeToo Go Back to Work." *The Atlantic*, October 12. Retrieved from https://www.theatlantic.com/entertainment/archive/2018/10/has-metoo-actually-changed-hollywood/572815/.

Gilbert, S. (2018b). "*The Romanoffs* Defends the Men of #MeToo." *The Atlantic*, November 8. Retrieved from https://www.theatlantic.com/entertainment/archive/2018/11/romanoffs-defends-men-metoo/575356/.

Gilbert, S. (2020). "*The Assistant* and the Messes Women Clean Up." *The Atlantic*, February 6. Retrieved from https://www.theatlantic.com/culture/archive/2020/02/assistant-kitty-green-complicity-cleanup/606067/.

Gonzalez, S., and Friedlander, W. (2019). "'SNL' Fires New Hire Shane Gillis." *CNN*, September 17. Retrieved from https://www.cnn.com/2019/09/16/entertainment/snl-shane-gillis/index.html.

Graham, R. (2017). "The 'Weinstein Effect' Hits a Wall." *The Boston Globe*, November 28. Retrieved from https://www.bostonglobe.com/opinion/2017/11/28/the-sexual-harassment-reckoning-takes-turn-the-weinstein-effect-hits-wall/oaFBb6caAFUjD7sC1Tt6GO/story.html.

Gray, E. (2019). "Gretchen Carlson Can't Talk About 'The Loudest Voice,' Which Is Why She Hopes Others Will." *Huffington*

Post, July 8. Retrieved from https:// www.huffpost.com/entry/gretchen-carlson-roger-ailes-the-loudest-voice_n_5d237844e4b04c481418283e.

Green, J., and Sakoui, A. (2019). "Men of #MeToo Are Back and No One Knows Quite How to Respond." *Bloomberg Wire Service*, February 27.

Gupta, A. (2019). "What's the Right Recompense?" *New York Times*, December 18. A2(L).

Guthrie, M. (2019). "NBCUniversal Says Women Are Not Silenced by NDAs, but Payout Terms Stay Secret." *The Hollywood Reporter*, October 29. Retrieved from https://www.hollywoodreporter. com/news/nbc-news-says-women-are-not-silenced-by-ndas-but-payout-terms-stay-secret-1250924.

Hall, S., ed. 1997. *Representation: Cultural Representations and Signifying Practices*. Vol. 2. Sage.

Hallemann, C. (2018). "Christina Hendricks Reflects on Her *Romanoffs'* Character's Shocking Death." *Town & Country*, October 20. Retrieved from https://www.townandcountrymag. com/leisure/arts-and-culture/ a23899404/christina-hendricks-interview-romanoffs-episode-3-house-of-special-purpose-death/.

Halloway, D. (2018). "Netflix Wants Aziz Ansari's 'Master of None' to Return for Season 3, Originals Chief Says." *Variety*, July 29. Retrieved from https://variety. com/2018/tv/news/netflix-aziz-ansari-master-of-none-1202889434/.

Haring, B. (2019). "#MeToo: Ex-Fox News Employees Ask for NDAs to Be Lifted in Wake of NBCUniversal Decision." *Deadline*, October 28. https://deadline. com/2019/10/ex-fox-news-employees-ask-for-non-disclosure-agreements-to-be-lifted-in-wake-of-nbcuniversal-decision-1202771283/#

Hild, A. (2018). "MeToo: The Reactions in Europe." October 27. *Young Feminist Europe*. Retrieved from https:// www.youngfeminist.eu/2018/02/ metoo-reactions-europe/https://www. theguardian.com/world/2018/oct/08/ metoo-one-year-on-hollywood-reaction.

Horan, M. (2016). "Grey's Anatomy Has Become Torture Porn." *Refinery29*, January 12.

Hutchinson, P. (2018). "#MeToo and Hollywood: What's Changed in the Industry a Year On?" *The Guardian*, October 8. Retrieved from https://www.refinery29.com/en-us/2016/01/101026/greys-anatomy-meredith-attacked.

Jarvy, N. (2017). "#MeToo Movement Tops 1.7 Million Tweets." *The Hollywood Reporter*, October 24. Retrieved from https://www.hollywoodreporter. com/news/metoo-%20movement-tops-17-million-tweets-1051517.

Jasheway, L. (2018). "Comedy Writers and Satirists of the #MeToo Movement." *Writer's Digest*, May 25. Retrieved from https://www.writersdigest.com/write-better-nonfiction/comedy-writers-satirists-metoo-movement.

Jenkins, H. (2019). "Apple Joins the Streaming Melange; Profitable Days Are Ahead for Hollywood Types Who Can Survive the #MeToo Purge." *Wall Street Journal*, April 2. Retrieved from https://www.wsj.com/ articles/apple-joins-the-streaming-melange-11554246402.

Jha, N. (2018). "Bollywood Needs a Time's Up Movement. Here's Why It's Not Happening Anytime Soon." *Buzz-Feed News*, November 3. Retrieved from https://www.buzzfeednews.com/ article/nishitajha/bollywood-metoo.

Judge, M. (2018). "Sexual Assault Case Against Anthony Anderson Dropped by L.A. DA." *The Root*, September 4. Retrieved from https:// thegrapevine.theroot.com/los-angeles-district-attorney-will-not-pursue-sexual-as-1828814996.

Kaminsky, M. (2019). "Five Biggest Sexual Harassment Cases." *Legalzoom*. Retrieved from https://www. legalzoom.com/articles/five-biggest-sexual-harassment-cases.

Kang, I. (2019). "MeToo-Inspired Stories Have Been Making Women Villains as Well as Victims." *Slate*, July 8. Retrieved from https://slate.com/ culture/2019/07/big-little-lies-late-night-metoo-stories-women-abusers-villains.html.

Kantor, J. (2017). "Tarantino on Weinstein: 'I Knew Enough to Do More Than I Did.'" *New York Times*, October 19. Retrieved from https://

www.nytimes.com/2017/10/19/movies/tarantino-weinstein.html?auth=login-email&login=email.

Kantor, J., and Twohey, M. (2017). "Harvey Weinstein Paid Off Sexual Harassment Accusers for Decades." *New York Times,* October 17. Retrieved from https://www.nytimes.com/2017/10/05/us/harvey-weinstein-harassment-allegations.html.

Karlis, N. (2018). "Bill Mahar Asks Ronan Farrow Has MeToo Gone Too Far?" *Salon,* April 28. Retrieved from https://www.salon.com/2018/04/28/bill-mahar-asks-ronan-farrow-has-metoo-gone-too-far/.

Kelly, S. (2019). "How *Black Christmas* Became a Slasher Movie for the "MeToo Era." *Los Angeles Times,* December 11. Retrieved from https://www.latimes.com/entertainment-arts/movies/story/2019-12-11/black-christmas-feminist-horror-sophia-takal.

Khorasani, S. (2019). "Harvey of Hollywood: The Face That Launched a Thousand Stories." *Hastings Communications and Law Entertainment Journal* 41, no. 1 (2019). Available At: https://repository.uchastings.edu/hastings_comm_ent_law_journal/vol41/iss1/5.

Kilkenny, K. (2019). "Brian Cox Talks 'Succession' Patriarch's Violence and 'Demonic' Drive." *The Hollywood Reporter,* September 15. Retrieved from https://www.hollywoodreporter.com/live-feed/succession-argestes-interview-brian-cox-1239650.

Kim, E. (2017). "Bill O'Reilly Speaks Out in First TV Interview Since Firing: 'This was a hit job.'" *The Today Show,* September 19. Retrieved from https://www.today.com/news/bill-o-reilly-sexual-harassment-allegations-against-him-was-hit-t116460.

Kirkland, J. (2019). "*Brooklyn Nine-Nine* Just Had One of the Most Realistic #MeToo Conversations on TV." *Esquire,* Mar 1. Retrieved from https://www.esquire.com/entertainment/tv/a26555109/brooklyn-nine-nine-me-too-episode-explained/.

Kornhaber, S. The Atlantic Culture Desk. (2019). "The 7 Most Defining #MeToo Moments of 2019." *The Atlantic,* December 12. Retrieved from https://

www.theatlantic.com/entertainment/archive/2019/12/most-defining-metoo-moments-2019/603490/.

Landsbaum, C. (2019). "'It was a pretty jaw-dropping moment': Jodi Kantor and Megan Twohey on Weinstein, Epstein, and the Future of #MeToo." *Vanity Fair,* September 10. Retrieved from https://www.vanityfair.com/news/2019/09/jodi-kantor-megan-twohey-she-said-weinstein-epstein-future-of-metoo.

Lanser, S. 1992. *Fictions of Authority: Women Writers and Narrative Voice.* Cornell University Press.

Lauzen, M. (2019a). "The Celluloid Ceiling: Behind-the-Scenes Employment of Women on the Top 100, 250, and 500 Films of 2018." Retrieved from https://womenintvfilm.sdsu.edu/wp-content/uploads/2019/01/2018_Celluloid_Ceiling_Report.pdf.

Lauzen, M. (2019b). "It's a Man's (Celluloid) World: Portrayals of Female Characters in the Top Grossing Films of 2018." Retrieved from https://womenintvfilm.sdsu.edu/wp-content/uploads/2019/02/2018_Its_a_Mans_Celluloid_World_Report.pdf.

Lawler, K. (2018). "In This #MeToo Era, It's Time to Retire Dating Shows for Good." *USA Today,* August 6. Retrieved from https://www.usatoday.com/story/life/tv/2018/08/06/era-me-too-retire-dating-shows-bachelorette/894993002/.

Lawler, K. (2018). "Will #MeToo Finally Make Us Question Violence Against Women on TV?" *USA Today,* October 23. Retrieved from https://www.usatoday.com/story/life/tv/2018/10/23/metoo-tv-shows-romanoffs-violence-against-women-matthew-weiner-christina-hendricks/1688357002/.

Lee, C. (2019). "*Bombshell* Director Says Minidress Scene Is Most Excruciating Thing He's Ever Filmed." *Vulture,* December 16. Retrieved from https://www.vulture.com/2019/12/inside-bombshells-excruciating-mini-dress-scene.html.

Lemiski, M. (2019). "TV Shows Are Navigating the #MeToo Movement Surprisingly Well." *Vice,* March 27. Retrieved from https://www.vice.com/en_us/

article/kzdv3z/tv-shows-are-navigating-the-metoo-movement-surprisingly-well.

Levine, N. (2018). "Nicole Kidman Gets Real About *Big Little Lies & #MeToo*." *Refinery29*, December 2. https://www.refinery29.com/en-gb/2018/12/218333/nicole-kidman-me-too-interview.

Lindsay, K. (2017). "*Broad City* Season 4, Episode 8 Recap: 'House-Sitting.'" *Refinery29*, November 15. Retrieved from https://www.refinery29.com/en-us/2017/11/181082/broad-city-season-4-episode-8-recap.

MacKinnon, C. (2019). "Where #MeToo Came From, and Where It's Going." *The Atlantic*, March 24. Retrieved from https://www.theatlantic.com/ideas/archive/2019/03/catharine-mackinnon-what-metoo-has-changed/585313/.

Main, A. (2017). "The #MeToo Hashtag Was Used in an Enormous Number of Tweets." *Mashable*, October 16. Retrieved from https://mashable.com/2017/10/16/me-too-hashtag- popularity/#m.6_OeIYYiq9.

Maple, T. (2019). "The 'Veep' #MeToo Episode Never Turned Real-Life Survivors Into a Punchline." *Bustle*, April 7. Retrieved from https://www.bustle.com/p/the-veep-metoo-episode-never-turned-real-life-survivors-into-a-punchline-17008980.

Marghitu, S. (2018). "It's just art': *Auteur* Apologism in the Post-Weinstein Era." *Feminist Media Studies*, 18:3, 491–494, DOI: 10.1080/14680777.2018.1456158 https://www.tandfonline.com/doi/ref/10.1080/14680777.2018.1456158?scroll=top.

Marshall, A. (2020). "Actors Walk Out After Roman Polanski Wins Best Director at France's Oscars." *New York Times*, February 28. Retrieved from https://www.nytimes.com/2020/02/28/movies/roman-polanski-cesar-awards-france.html.

McDonald, S. (2018). "After 'Get Out' and #MeToo, Steven Soderbergh's 'Unsane' Is All Too Unnerving." *The Undefeated*, March 22. Retrieved from https://theundefeated.com/features/unsane-steven-soderbergh-after-get-out-and-metoo-all-too-unnerving/.

McHenry, J. (2018). "How *The Good Fight* Brought the Trump Era to TV, from the Pee Tape to #MeToo." *Vulture*, June 19. Retrieved from https://www.vulture.com/2018/06/the-good-fight-season-2-pee-tape-me-too-donald-trump.htmlultur.

Merkin, D. (2018). "Publicly, We Say #MeToo. Privately, We Have Misgivings." *New York Times*, January 5. Retrieved from https://www.nytimes.com/2018/01/05/opinion/golden-globes-metoo.html.

Miller, J. (2018). "*Glow* Didn't Need Harvey Weinstein to Inspire Its #Me Too Moment." *Vanity Fair*, June 29. Retrieved from https://www.vanityfair.com/hollywood/2018/06/glow-netflix-season-2-me-too.

Miller, L. (2017). "'*The Good Fight*' Review: CBS Makes the Case for Its Netflix-Style Service With Emmy-Certified Spin-Off." *IndieWire*, February 17. Retrieved from https://www.indiewire.com/2017/02/the-good-fight-review-cbs-all-access-christine-baranski-1201784345/.

Miller, L. (2018). "'Murphy Brown' Gets a #MeToo-Themed Episode and Reclaims Her Power with '#MurphyToo.'" *IndieWire*. October 11. Retrieved from https://www.indiewire.com/2018/10/murphy-brown-murphytoo-review-candace-bergen-metoo-1202011581/.

Miller, L. (2019). "How 'The Loudest Voice' Portrays Gretchen Carlson as Its Real Hero." *The Hollywood Reporter*. August 4. Retrieved from https://www.hollywoodreporter.com/live-feed/loudest-voice-episode-6-explained-gretchen-carlsons-lawsuit-1228948.

Miller, T., and Armstrong, K. (2015). "An Unbelievable Story of Rape." *Propublica and the Marshall Project*, December 16. Retrieved from https://www.propublica.org/article/false-rape-accusations-an-unbelievable-story.

Moeggenberg, Z. and Solomon, S. (2018). "Power, Consent, and the Body: #MeToo and the Handmaid's Tale." *Gender Forum*, no. 70 (2018): 4–25.

Morris, R. (2018). "Is #MeToo Changing Hollywood?" *BBC News*, March 3. Retrieved from https://www.bbc.com/news/world-us-canada-43219531.

Nemzoff, R. (2017). "The Weinstein

Effect: Avalanche of Allegations Usher in a New Era." *Huffington Post*, December 20. Retrieved from https://www.huffingtonpost.com/entry/the-weinstein-effect-avalanche-of-allegations-usher_us_5a3ad40ee4b06cd2bd03d790.

The New Yorker. (2019). "Dream Hampton on 'Surviving R. Kelly,' Sexual Abuse, and Race." *The New Yorker*, January 22. Retrieved from https://www.newyorker.com/news/the-new-yorker-interview/dream-hampton-on-surviving-r-kelly-and-necessary-conversations-about-sexual-abuse-and-race.

New Yorker Radio Hour. (2019). "Emily Nussbaum on TV's 'Deluge' of #MeToo Plots, May 31." Retrieved from https://www.stitcher.com/podcast/wnyc/the-new-yorker-radio-hour/e/61532698.

Newis-Smith, J. (2019). "Fleabag's Phoebe Waller-Bridge on How She Expertly Navigates Sexual Jokes in a Post #MeToo World." *Glamour UK*, April 2. Retrieved from https://www.glamourmagazine.co.uk/article/phoebe-waller-bridge-chewbacca-star-wars-interview.

Nicolaou, E. (2019). "Laurie Luhn & Roger Ailes' Relationship Was as Complicated in Real Life as It Is in *The Loudest Voice*." *Refinery29*, July 14. Retrieved from https://www.refinery29.com/en-us/2019/07/237533/where-is-laurie-luhn-now-loudest-voice-roger-ailes-relationship.

Nicolaou, E., and Smith, C. (2019). "A #MeToo Timeline to Show How Far We've Come and How Far We Need to Go." *Refinery29*, October 7. Retrieved from https://www.refinery29.com/en-ca/2019/10/8534374/a-metoo-timeline-to-show-how-far-weve-come-how-far-we-need-to-go?utm_source=email&utm_medium=email_share.

Nissen, D. (2019). "Woody Allen: 'I've done everything the #MeToo movement would love to achieve.'" *Variety*, September 6. Retrieved from https://variety.com/2019/film/news/woody-allen-metoo-1203326247/#!

Nordine, M. (2018a). "Liam Neeson Fears the #MeToo Movement Is Becoming a 'Witch Hunt.'" *IndieWire*, January 13. Retrieved from https://www.indiewire.com/2018/01/liam-neeson-dustin-hoffman-sexual-harassment-1201917127/.

Nordine, M. (2018b). "Michael Haneke Criticizes #MeToo: 'Witch hunts should be left in the Middle Ages.'" *IndieWire*, February 11.

Nuraddin, N. (2018). "The Representation of the #Metoo Movement in Mainstream International Media." Retrieved from http://urn.kb.se/resolve?urn=urn:nbn:se:hj:diva-40839.

Nussbaum, E. (2016). "'Fleabag,' an Original Bad-Girl Comedy." *The New Yorker*, September 26. Retrieved from https://www.newyorker.com/magazine/2016/09/26/fleabag-an-original-bad-girl-comedy.

Nussbaum, E. (2019). "TV's Reckoning with #MeToo." *The New Yorker*, May 27. Retrieved from https://www.newyorker.com/magazine/2019/06/03/tvs-reckoning-with-metoo.

O'Malley, S. (2020). "The Assistant" (review). *RogerEbert.com*, January 31. Retrieved from https://www.rogerebert.com/reviews/the-assistant-movie-review-2020.

Orgad, S. (2014). *Media Representation and the Global Imagination*. John Wiley & Sons.

Oulaw, K. (2019). "The Boys: How Harvey Weinstein and #MeToo Changed the Show's Treatment of a Controversial Comic Scene." *Comicbook*, July 31. Retrieved from https://comicbook.com/comics/2019/07/31/the-boys-tv-series-starlight-rape-scene-comic-changes/.

Page, N. (2019). "How New #MeToo Laws in New York and California Could Change Film and TV Production." *IndieWire*, May 29. Retrieved from https://www.indiewire.com/2019/05/anti-harassment-laws-new-york-california-effect-film-tv-production-1202145394/.

PBS. 2018. "Me Too, Now What?" *PBS*, 2018. Retrieved from https://www.pbs.org/show/metoo-now-what/.

Peltier, E. (2020). "Adele Haenel: France 'Missed the Boat' on #MeToo." *New York Times*, February 24. Retrieved from https://www.nytimes.com/2020/02/24/movies/adele-haenel-france-metoo.html.

Perman. S. (2019). "John Lasseter Is Attempting Hollywood's Biggest #MeToo Comeback. How's That Going?" Tribune Interactive, LLC, last modified Jan 27.

Piacenza, J. 2018. "How #MeToo Impacts Viewer' Decisions on What to Watch." *Morning Consult,* May 28, 2018. Retrieved from https://morningconsult. com/2018/05/28/how-metoo-impacts-viewers-decisions-what-watch/.

Pompeo, J. (2019). "'Jesus, are we doing this again?' At *Today,* a 'Wound being opened' as Ronan Farrow Takes Aim at Laurer, NBC Management." *Vanity Fair,* October 10. Retrieved from https://www.vanityfair.com/news/2019/10/ronan-farrow-takes-aim-at-matt-lauer-nbc-management.

Poniewozick, J. (2019). "How 'Survivor' Failed Its #MeToo Test." *New York Times,* December 12. Retrieved from https://www.nytimes.com/2019/12/12/arts/television/survivor-dan-spilo.html.

Prange, S. (2018). "Top Film School Deans Use #MeToo Scandal as Teachable Moment." *Variety* 339 (4): 10. 10.

Prasad, V. (2018). "If Anyone Is Listening, #MeToo: Breaking the Culture of Silence Around Sexual Abuse Through Regulating Non- Disclosure Agreements and Secret Settlements." 59 *Boston College Law Review* 2507 (2018). Retrieved from https://lawdigitalcommons.bc.edu/bclr/vol59/iss7/8.

Ravindran, M. (2018). "'*Fleabag*' Creator Phoebe Waller-Bridge: Viewers Crave More Female-centric Stories." *Screen Daily,* April 10. Retrieved from https://www.screendaily.com/news/fleabag-creator-phoebe-waller-bridge-viewers-crave-more-female-centric-stories/5128142.article.

Reddit Thread. (2017). "Dr. Melfi Rapist Never Punished?" *Reddit.* Retrieved from https://www.reddit.com/r/thesopranos/comments/4wnswl/dr_melfi_rapist_never_punished/.

Rice, L. (2017). "The Future (of TV) Is Female." *Entertainment Weekly* (1558/1559), 10–11.

Richwine, L. (2018). "After #MeToo, Hollywood Women Seize Power Behind TV Camera." *USA News,* August 16. Retrieved from http://usa-news.org/2018/08/16/after-metoo-hollywood-women-seize-power-behind-tv-camera/.

Rolling Stone. (2018). "Watch Jamie Lee Curtis Talk 'Halloween,' Trauma in Conversation at 92Y." *Rolling Stone,* October 8. Retrieved from https://www.rollingstone.com/movies/movie-news/jamie-lee-curtis-halloween-trauma-metoo-92y-734381/.

Romano, R. (2017). "*Broad City* Star Ilana Glazer Says She Fired Staffers After Being Sexually Harassed." *Entertainment Weekly,* October 18. Retrieved from https://ew.com/tv/2017/10/18/broad-city-ilana-glazer-metoo-sexual-harassment/.

Roxborough, S. (2019). "Oscars: International Film Race Puts Spotlight on Extreme Female Characters." *The Hollywood Reporter,* November 18. Retrieved from https://www.hollywoodreporter.com/news/oscars-international-race-puts-spotlight-extreme-female-characters-1252013.

Roxborough, S., and Richford, R. (2018). "Europe's Bitter #MeToo Debate: Bardot, Bertolucci and the 'Threat of Change.'" *The Hollywood Reporter,* May 3. Retrieved from https://www.hollywoodreporter.com/features/europes-bitter-metoo-debate-bardot-bertolucci-threat-change-1107198.

Rubin, A., and Peltier, E. "In France, Accusation Plays Out Differently." *New York Times,* July 31. Retrieved from https://search-ebscohost-com.library.esc.edu/login.aspx?direct=true&db=a9h&AN=130981610&site=ehost-live.

Rugendyke, L. (2018). "Sarah Snook on Teenage Rebellion, #MeToo and Taking on HBO's *Succession.*" *The Sydney Morning Herald,* May 31. Retrieved from https://www.smh.com.au/entertainment/sarah-snook-on-teenage-rebellion-metoo-and-taking-on-hbos-succession-20180531-h10rz6.html.

Salmon, C. (2018). "How TV Comedies Are Leading the Way on #MeToo." *Little White Lies,* October 4. Retrieved from https://lwlies.com/articles/tv-comedies-metoo-bojack-horseman-glow-kimmy-schmidt/.

Sandberg, B., and Masters, K. (2019). "'The

environment was very toxic': Nudity, a Graphic Photo and the Untold Story of Why Ruth Wilson Left 'The Affair.'" *The Hollywood Reporter*, December 18, 2019. Retrieved from https://www.hollywoodreporter.com/features/ruth-wilson-left-affair-hostile-environment-nudity-issues-1263553.

Sarachan, R. 2019. "Rob Delaney and Sharon Horgan Discuss the Fourth and Final Season of Amazon's 'Catastrophe.'" *Forbes*, March 27. Retrieved from https://www.forbes.com/sites/risasarachan/2019/03/27/rob-delaney-and-sharon-horgan-discuss-the-fourth-and-final-season-of-amazons-catastrophe/#48c4a3d57931.

Schaffstall, K. (2019). "Samantha Bee Hosts #MeToo-Themed Thanksgiving Dinner." *The Hollywood Reporter*, November 21. Retrived from https://www.hollywoodreporter.com/news/samantha-bee-throws-metoo-thanksgiving-dinner-survivors-1256869.

Schama, C. (2020). "*The Assistant* Is a Compact, Quiet #MeToo Movie with a Loud, Expansive Effect." *Vogue*, January 31. Retrieved from https://www.vogue.com/article/kitty-green-interview-the-assistant-movie.

Schwartzel, E., and Watson R. (2019). "Life and Arts: After MeToo, Films Face Test—New Movies Scrutinized Directors Show in Theaters Overseas, but Maybe Not in the U.S." *The Wall Street Journal*, August 26th.

Scott, M. (2019). "Busan: Mike Figgis to Direct Three Short Films in Korea About #MeToo Movement." *The Hollywood Reporter*, October 5. Retrieved from https://www.hollywoodreporter.com/news/mike-figgis-direct-short-films-korea-metoo-movement-1245660.

Serjeant, J. (2020). "After Weinstein, #MeToo Themes in Film, TV Reflect Wider Cultural Reckoning." *Reuters*, March 12. Retrieved from https://www.reuters.com/article/us-people-harvey-weinstein-culture-idUSKBN20Z1AG

Sharf, Z. (2018a). "Roman Polanski Bashes #MeToo : It's 'Collective Hysteria' and 'Total Hypocrisy.'" *IndieWire*, May 9. Retrieved from https://www.indiewire.com/2018/05/roman-polanski-metoo-hysteria-hypocrisy-1201962230/.

Sharf, Z. (2018b). "Sean Penn Criticizes #MeToo Movement for Dividing Men and Women, Being 'Too Black and White.'" *IndieWire*, September 17. Retrieved from https://www.indiewire.com/2018/09/sean-penn-metoo-divides-men-women-1202004196/.

Sharf, Z. (2019a). "Olivia Munn Criticizes Quentin Tarantino for 'Pushing Past Abusive Behavior' Without 'Earning' It." *IndieWire*, July 2. Retrieved from https://www.indiewire.com/2019/07/olivia-munn-quentin-tarantino-push-past-abuse-1202155195/.

Sharf, Z. (2019b). "Casey Affleck Speaks Out: 'It Scares Me' to Talk About #MeToo Movement." *IndieWire*, August 6. Retrieved from https://www.indiewire.com/2019/08/casey-affleck-scared-talk-metoo-movement-harassment-1202163605/.

Shaw, L. (2019). "Anita Hill Takes on Hollywood's #MeToo Culture with Huge Industry Survey." *Bloomberg*, November 20. Retrieved from https://www.bloomberg.com/news/articles/2019-11-20/anita-hill-takes-on-hollywood-s-metoo-culture-with-huge-survey.

Siapera, E. (2010). *Cultural Diversity and Global Media: The Mediation of Difference.* John Wiley & Sons.

Siegel, T. (2018). "#MeToo Hits Movie Deals: Studios Race to Add 'Morality Clauses' to Contracts." *The Hollywood Reporter*, February 7. Retrieved from https://www.hollywoodreporter.com/news/metoo-hits-movie-deals-studios-race-add-morality-clauses-contracts-1082563.

Sims, D. (2020). "*The Assistant* is a Subtle Film for the #MeToo Era." *The Atlantic*, February 1. Retrieved from https://www.theatlantic.com/culture/archive/2020/02/assistant-review-julia-garner-harvey-weinstein/605875/.

Smith, C. (2018). "To All the Boys I've Loved Before Updates the Romantic Comedy for the #MeToo Era." *PopMatters*, October 1, 2018. Retrieved from https://www.popmatters.com/all-boys-ive-loved-before-2608644567.html?rebelltitem=1#rebelltitem1.

Specter, E. (2019). "*Promising Young Woman* Is the Revenge Movie the #MeToo Era Deserves." *Vogue*, December 12. Retrieved from https://www.

vogue.com/article/promising-young-woman-trailer.

Stewart-Kroeker, S. (2019). "'What Do We Do with the Art of Monstrous Men? Betrayal and the Feminist Ethics of Aesthetic Involvement." *De Ethica. a Journal of Philosophical, Theological and Applied Ethics* (2019): 1–24.

Strause, J. (2019). "'*Veep*' Boss on Tackling #MeToo and the Biggest Threat Ahead." *The Hollywood Reporter,* April 7. Retrieved from https://www.hollywoodreporter.com/live-feed/veep-tackles-metoo-jonah-ryan-david-mandel-interview-1199294.

Thom, M. (2018). "Now #MeToo Has Its Own Horror Movie." *Spiked,* October 31. Retrieved from https://www.spiked-online.com/2018/10/31/now-metoo-has-its-own-horror-movie/.

Thompson, R. (2019). "Emma Watson and Time's Up Launch Free Hotline for Women Experiencing Workplace Harassment." *Mashable,* August 5. Retrieved from https://mashable.com/article/emma-watson-times-up-uk-advice-line/.

Tolentino, J. (2030). "The Opening Statements in the Harvey Weinstein Trial and the Undermining of 'MeToo,'" *The New Yorker,* January 23. Retrieved from https://www.newyorker.com/news/dispatch/the-opening-statements-in-the-harvey-weinstein-trial-and-the-undermining-of-metoo?utm_source=onsite-share&utm_medium=email&utm_campaign=onsite-share&utm_brand=the-new-yorker.

Traister, R. (2018). *Good and Mad: The Revolutionary Power of Women's Anger.* Simon & Schuster.

Traister, R. (2019). "The Toll of Me Too." *The Cut.* September 30. Retrieved from https://www.thecut.com/2019/09/the-toll-of-me-too.html.

Travers, P. (2018). "'*Halloween*' Review: A Slasher Movie Reboot for the #MeToo Era." *Rolling Stone,* October 16. Retrieved from https://www.rollingstone.com/movies/movie-reviews/halloween-movie-review-737665/.

Turchiano, D. (2018). "'Jane the Virgin' Boss Breaks Down Revisiting the Jane-Chavez Relationship Through the 'Me Too' Lens." *Variety,* March 2. Retrieved from https://variety.com/2018/tv/news/jane-the-virgin-jennie-urman-me-too-times-up-power-dynamics-1202714702/.

Turchiano, D. (2019). "'Unbelievable' Bosses on Adapting an Unreliable Witness' Assault Story." *Variety,* September 13. Retrieved from https://variety.com/2019/tv/features/unbelievable-marie-netflix-susannah-grant-sarah-timberman-interview-1203305390/.

Vance, C. (2020). "Harvey Weinstein Sentenced to 23 Years in Prison." Press Release of the Manhattan District Attorney's Office, March 11. Retrieved from https://www.manhattanda.org/d-a-vance-harvey-weinstein-sentenced-to-23-years-in-prison/.

Verongos, H. (2019). "'Untouchable' Review: A Documentary of the Harvey Weinstein Case." *New York Times,* September 2. Retrieved from https://www.nytimes.com/2019/09/02/movies/untouchable-review.html.

Villarreal, Y. (2018). "In 2018, Prominent TV Shows Took a Stand on "#MeToo and Its Aftermath." *Los Angeles Times,* December 21. Retrieved from https://www.latimes.com/entertainment/tv/la-et-st-metoo-episodes-tv-2018-story.html.

Vox. (2019). "A List of People Who Are Accused of Sexual Assault Since 2017." *Vox,* January 19. Retrieved from https://www.vox.com/a/sexual-harassment-assault-allegations-list/frankie-shaw.

Vulture Editors. (2018). "Elizabeth Moss, Tyra Banks, and More on What Has Changed in Hollywood Post-#MeToo." *Vulture,* October 21. Retrieved from https://www.vulture.com/2018/10/metoo-hollywood-one-year-later.html.

Walsh, K. (2019). "Charlize Theron's *Bombshell* Transformation Wows at First Los Angeles Screening." *Vanity Fair,* October 14. Retrieved from https://www.vanityfair.com/hollywood/2019/10/bombshell-charlize-theron-nicole-kidman-margot-robbie?utm_source=onsite-share&utm_medium=email&utm_campaign=-onsite-share&utm_brand=vanity-fair.

Wandsworth, N. (2017). "Louis C.K.'s Apology Is Imperfect. but It Is Still Important." *Washington Post,* November 14. Retrieved from https://www.washingtonpost.com/

news/monkey-cage/wp/2017/11/14/louis-c-k-s-apology-was-imperfect-but-it-was-still-important/?utm_term=.fa6734241eb3.

Watercutter, A. (2018). "Film Festivals Are Forever Changed in the Wake of #MeToo." *Wired*, January 29. Retrieved from https://www.wired.com/story/film-festivals-me-too-era/.

Weinstein, H. (2017). "Statement from Harvey Weinstein." *New York Times*, October 5. Retrieved from https://www.nytimes.com/interactive/2017/10/05/us/statement-from-harvey-weinstein.html.

Weiss, B. (2018). "Aziz Ansari Is Guilty. Of Not Being a Mind Reader." *New York Times*, January 15. Retrieved from https://www.nytimes.com/2018/01/15/opinion/aziz-ansari-babe-sexual-harassment.html?auth=login-email&login=email.

Wilkinson, A. (2018). "Unbreakable Kimmy Schmidt Deftly Take on #MeToo and Terrible Men in Season 4." *Vox*, May 30. Retrieved from https://www.vox.com/culture/2018/5/30/17391290/unbreakable-kimmy-schmidt-season-4-review-netflix-spoilers-metoo.

Wilstein, M. (2018). "Bill Maher Warns #MeToo Movement Creating 'Police State' for Sex. *The Daily Beast*, February 10. Retrieved from https://www.thedailybeast.com/bill-maher-warns-metoo-movement-creating-police-state-for-sex.

Winfrey, O. (2018). "Oprah Winfrey Receives the Cecil B. DeMille Award." *Golden Globe*. Retrieved from https://youtu.be/tnUQKxPWS9U

Wolper, C. (2018). "The #MeToo Moment Everyone's Missing Totally Changes How We Think About *BoJack Horseman*." *Slate*, October 2. Retrieved from https://slate.com/culture/2018/10/bojack-horseman-metoo-moment-overlooked.html.

Zacharek, S. (2018). "First, the Movies Now, the World: How Corrupt, Craven Hollywood Found Its Conscience." *Time International* (Atlantic Edition), March 12, 2018.

Zacharek, S., Dockterman, E., and Edwards, H. (2017). "The Silence Breakers." *Time Magazine*, December 18. Retrieved from https://time.com/time-person-of-the-year-2017-silence-breakers/.

Zeitchik, S. (2019). "A Star's Letter and the Future of #MeToo." *Washington Post*, March 3.

Zhang, J. (2018). "#MeToo Has Reached China, but Will It Have an Impact?" *The Hollywood Reporter*, January 8. Retrieved from https://www.hollywoodreporter.com/news/metoo-has-reached-china-but-will-it-have-an-impact-1071819.

Zimmerman, A. (2019a). "How SAGE Is Complicit in Hollywood's Casting Couch Culture." *The Daily Beast*, February 10. https://www.thedailybeast.com/inside-hollywoods-abusive-casting-couch-culture.

Zimmerman, A. (2019b). "Max Landis' Sexual-Assault Accuser Speaks Out Against His Hollywood 'Comeback': The Anonymous Woman Spoke to the Daily Beast About the Terrifying Encounter in 2012—Which Landis Admitted Was 'a Boundaries Violation'—and Why She's Decided to Share Her Story Now." *The Daily Beast*, February 13. Retrieved from http://library.esc.edu/login?url=https://search.proquest.com/docview/2179165781?accountid=8067.

Index

191